BABYLON'S BANKSTERS

THE ALCHEMY OF DEEP PHYSICS,
HIGH FINANCE AND ANCIENT RELIGION

AN ESSAY CONCERNING THE RELATIONSHIPS
BETWEEN AETHER PHYSICS, ECONOMICS,
ASTROLOGY, ALCHEMY, GEOMANCY, ANCIENT
TEMPLES, AND THE POLITICS OF SUPPRESSION

JOSEPH P. FARRELL

FERAL HOUSE

BABYLON'S BANKSTERS

"Alas, alas, that great city Babylon, that mighty city!
for in one hour is thy judgment come.
And the merchants of the earth shall weep and
mourn over her; for no man buyeth their
merchandise any more:
The merchandise of gold, and silver, and precious stones,
and of pearls, and fine linen, and purple, and silk, and
scarlet, and all thine wood, and all manner vessels of
ivory, and all manner vessels of most precious wood,
and of brass, and iron, and marble,
And cinnamon, and odours, and ointments, and
frankincense, and wine, and oil, and fine flour, and
wheat, and beasts, and sheep, and horses, and chariots,
and slaves, and souls of men.
And the fruits that thy soul lusted after are departed
from thee, and all things which were dainty and goodly
are departed from thee, and thou shalt find them
no more at all.
The merchants of these things, which were made rich
by her, shall stand afar off for the fear of her torment,
weeping and wailing,
And saying, Alas, alas, that great city, that was clothed
in fine linen, and purple, and scarlet, and decked with
gold, and precious stones, and pearls!
For in one hour so great riches is come to nought. And
every shipmaster, and all the company in ships, and
sailors, and as many as trade by sea, stood afar off.
And cried when they saw the smoke of her burning,
saying, What [city is] like unto this great city!
And they cast dust on their heads, and cried, weeping
and wailing, saying, Alas, alas, that great city, wherein
were made rich all that had ships in the sea by reason
of her costliness! for in one hour is she made desolate.

The Revelation of St. John
18: 10-19

Babylon's Banksters:
The Alchemy of Deep Physics, High
Finance and Ancient Religion
© 2010 by Joseph P. Farrell
All rights reserved.
A Feral House book.
ISBN 978-1-93259-579-6

Feral House
1240 W. Sims Way Suite 124
Port Townsend WA 98368
www.FeralHouse.com
Book design by Jacob Covey.

10 9 8 7 6 5 4 3 2 1

To
Scott Douglas deHart:
Anything I could say, any gratitude I could express,
is simply inadequate for you;

And to
Tracy S. Fisher:
You are, and will always be, sorely missed;

And to Richard C. Hoagland:
In thanks for so many wonderful insights, stimulating thoughts,
splendid conversations, and scintillating analysis;

And to
George Ann Hughes:
True and generous in your prayers and support;

He who has found such as these, has found treasures

ACKNOWLEDGEMENTS

As always, I'd like to thank Mr. Richard C. Hoagland for contributions to this book. In this case, unbeknownst to Mr. Hoagland, I had determined to write a book on this subject a couple of years ago, and in one of those synchronicities that seem to be a common feature of modern life, while listening to one of his presentations at a conference, a source for research, utilized and cited herein, was mentioned. Needless to say, I quickly jotted down the reference, and eventually, when I purchased the book, found it to be a goldmine of information and thought-provoking ideas. The book was astrologist Robert Gover's *Time and Money: The Economy and the Planets*, and notwithstanding the short review of it in the main text, it is in itself a book worth considering in its totality for those interested in the more esoteric aspects of economic forecasting and activity.

Mr. Hoagland also played another role in the formation of this book, and again, it was one that at the time he didn't realize he was playing. During a visit to his home in 2008, Mr. Hoagland showed me volumes of data in his library that had been collected and analyzed by the little-known Foundation for the Study of Cycles. We had some stimulating conversation over these materials, not the least of which were their implications for the connection between physics and finance. In due course I set off researching the availability of some of this material, and eventually was able to purchase the classic text *Cycles: The Science of Prediction* by the Foundation's founder, economist Edward Dewey.

In researching this book, I quickly discovered that the materials to document the main themes were so numerous and varied that any attempt at comprehensiveness was futile. Notwithstanding that, a word of thanks is due to all those inventors and scientists who have seen their discoveries and theories shuffled to the sidelines by a "corporautocracy" that is hell-bent on keeping humanity in the slavery of an energy and financial dark ages: to all the Teslas, Farnsworths, Kozyrevs, Müllers, Richters, DiPalmas, Bedinis, and Beardens out there: thank you. I only regret that I couldn't mention you all, but rest assured, your work and its implications *were* noticed.

And to all the other researchers out there who have contributed to unmasking the sinister role of private banking in the affairs of state and of mankind throughout history: thank you. Again, it was not for the dearth but the surfeit of information that specific mention could only be made of a few of you.

And with that, on to the story…

Joseph P. Farrell
Spearfish, South Dakota, 2009

TABLE OF CONTENTS

Acknowledgements	9

PREFACE
Prologue Is Epilogue, or, Two Flies in the Ointment: Communist China And Nazi Germany — 19

A. Confronting the Beast	19
B. Meltdown, or Message?	21
1. David Li's Formula	21
2. The Li Clan, the Canadian Imperial Bank of Commerce, and the Triads	27
3. China's Money	31
C. Nazi Germany: Physics and Finance Fully Rationalized	32
D. Energy and Money-Creating Autarchy	36

PART ONE
Historical and Conceptual Background — 43

1. The Conspirators' Postwar Détente	45
A. A Hypothetical Scenario	46
B. Estulin's Study and Standard Interpretations	50
1. Estulin's Sources	50
2. Protocols	51
a. Attendees	51
b. Vetting, Securing, and Announcing the Location	55
c. The Sessions and Their Rules	56
3. Goals and History: Bilderberger Shenanigans	57
4. The Standard Interpretive Temptation	60
C. Unusual Guests at the First Meeting and the Alternative Explanation: Détente	64
1. Two Prominent Guests with Nazi Backgrounds	64

2. The Wider Context of the Postwar Nazi International and the Name "Bilderberger" Group	65
3. An Event Indicative of an Understanding between the Anglo-American Corporate Globalist Elite and the Nazi International	67
4. Closed Systems, Open Systems, and Further Possible Reasons for the Détente	70
5. An Analogy of the Détente: The Nazi-Soviet Pact	73
6. Closed Systems, Globalization, New Energy Sources, and Outer Space	73
D. Conclusions and Implications	74
2. Hoover's Hidden Legacy and Gift	77
A. An Overview of Dewey's Database	79
B. The Inevitability and Predictability of Cycles: Closed vs. Open Systems	82
1. Criticism from the Conventional	88
C. Cycles, Trends, and Cycles Upon Cycles	90
1. The Period Patterns	90
2. "Overlays": Combined Cycles as Multiwave Modulation, and Possibly as Interferometry?	92
3. Inevitability, Predictability, and Human Actions	96
4. Waves, Wars, and Revolutions	97
5. Waves, Overlays, and Modulation: The Physics Analogies Employed by Dewey and Dakin, and Their Implications	98
D. Conclusions: Implications, and a Segue in the Form of Significant Questions	102
3. Germany, RCA, and J.P. Morgan: Cases of Interest, and Suppression	107
A. Dr. Hartmut Müller and Global Scaling Theory	107
1. Longitudinal Waves in the Physical Medium	110
2. The Link: Geometries	113
a. Planetary Alignments and Signal Propagation	114
b. The Electrically Dynamic Solar System and Planetary Alignments	119
c. Plasma Cosmology	119
d. Plasma Transduction of the Vacuum or Zero-Point Energy: Dr. Ronald Richter Revisited	124
e. Rotating Systems within Rotating Systems	127
B. All Roads in Physics Lead To Tesla	130
1. Colorado Springs	131
2. Wardenclyffe	136
a. Weaponizing Wardenclyffe and Suppressing the Scalars: Tesla at Tunguska and Morgan at Mischief	137
b. Lt. Col. Tom Bearden on Scalar Resonance	147
C. Conclusions	152

PART TWO:
THE TEMPLES, THE STARS, AND THE BANKSTERS 156

4. TEMPLES, TEMPLATES, AND TRUSTS: THE ANCIENT ROOTS 159
   OF A DEEP RELATIONSHIP
   A. Temples and Trusts 160
      1. The Roman Model 160
         a. The Bullion Trust and the Temple 161
         b. A Fascinating Tangent: Byzantium, Religion, ... 163
            and the Money Power
      2. The Egyptian Model: Mining, Slavery, Mercenaries, ... 165
         and Implications
         a. Nubia and Egypt 165
         b. Quartz, Gold, and Slavery 166
         c. Diodorus Siculus and the Bisharee Mines 167
         d. Money, Kings, Temples, and Counterfeiters 169
   B. Temples and Templates: Astronomy, Astrology, and the 171
      Alchemy of Money
      1. The Temples and the Stars 171
      2. The Peculiar Resemblance of Pantheons Between Cultures ... 175
      3. Gold, Gods, and Gems 176
         a. The Traditional Powers Associated with Gems 177
            (1) Special Case Number One: Invisibility 177
            (2) Special Case Number Two: Anti-Gravity 178
            (3) Special Case Number Three: Light Absorbing 179
                and Emitting Stones
         b. Gemstones, The Zodiac, and the Hebrew High Priest's ... 180
            Ephod, or Breastplate
   C. Conclusion 184

5. MONEY, MONOTHEISM, MONARCHIES, AND MILITARIES: 187
   THE THESIS OF DAVID ASTLE
   A. The State of Evidence and the Need for Speculation 187
      1. Ancient and Modern Banking Conspiracy 189
   B. The Medium of Exchange and Bullion as an Order on 189
      State Warehouses
      1. The Control of Mining and Bullion 191
      2. Ancient Babylon and Egypt 195
         a. Early Egypt's Independence from the Babylonian 195
            Money Power
      3. The Conspirators at Work 197
         a. Economic and Military Autarchy and Modern Analogues: ... 197
            Sparta and the Greek Tyrants
         b. Babylon, Persia, and Money Creation 200

C. How the Conspiracy Worked	202
1. The First Stage: Penetrate and Ally with the Temple	202
2. The Second Stage: Issue False Receipts	203
3. The Third Stage: Substitute Bullion for Letters of Credit as a Measure Against False Receipts	203
4. The Fourth Stage: Then Create a Facsimile of Money	204
D. The Parallel to the Aftermaths of The Cosmic War, and World War II: The Globalist Agenda and Money, Monarchies, Monotheism, and Militaries	204
6. Alchemy Upsets the Applecart: The Transmutative Medium and the Alchemy of the Stars and Banksters	207
A. Econophysics	208
1. Physicists Invade Finance: The Modern Model as a Key to the Paleoancient Past	208
2. Quantum Mechanics, Ancient Astrology, and the Statistical Approach	209
3. Dr. Li's Gaussian Copula Formula and a Physics Analogue: The Multibody Problem	210
4. The Deeper Physics	211
a. David Bohm's Hidden Variable Quantum Mechanics and the Implicate Order	211
b. The Foundation for the Study of Cycles Notices a Similar Thing	216
c. The Well-Tempered Clavier: The First Physical Unification	217
d. Nikolai Kozyrev's Causal Mechanics and Precursor Engineering	218
B. Economics, Astrology, and Astrophysics	220
C. Ellen Hodgson Brown	225
1. The Depression of the 1780s and the Banksters	225
2. The Depression of the 1870s and the Banksters	230
3. The Great Depression of the 1930s and the Banksters	234
4. Implications	236
D. Implications of Engineerability: The Ancient Alchemical Connection	238
PART THREE: THE MONSTERS IN THE MACHINE	243
7. Sacred Sites and Scalar Temples: The Earth Grid and the Transmutative Medium	245
A. The Modern Rise of Earth Grid Theories	247
1. Ivan Sanderson	247
2. The Russians Get Into the Game	248
3. Back to the Nazis	250

  B. Dr. Konstantin Meyl's Paleophysical Interpretation of Ancient   251
   Temples as Scalar Resonators

8. Templates, Genomes, and Banksters: Or, Why Do They All   267
  Marry Cousins and End Up With Colossally Stupid Kids?
  A. Ancient Rome   267
   1. The Change in Roman Racial Stock and Imperial Policy   271
   2. The Next Stage: Venice and Banking   272
   3. On to Amsterdam, London, the Reformation,   272
    and the Wars of Religion
  B. The Myth of the Rothschild Descent from Nimrod: A Second Look   274
  C. Human DNA and the Hermetic Code   275
   1. The I Ching   275
   2. The Well-Tempered Cosmos   279
  D. The Ancient Contact: The Rothschild Nimrod Myth   284
   in a Wider Context

9. The Banksters' Real Business: The Pattern of War, Scarcity,   287
  Suppression, Slavery and Monopolization
  A. The Historical Patterns of Suppression   288
   1. Of the Financial Alchemy of Money as Credit of the Nation   288
   2. Of the Physical Alchemy of Energy from the Medium   291
  B. The Physics, The Financial Alchemy, and the Banksters   292
   1. Assembling the Pieces   292
   2. The Underlying Principle   293
    a. The Reasons for the Banksters' Ancient Association   294
     with the Temple
   3. The Possible Agenda   295

Bibliography   300

Preface

PROLOGUE IS EPILOGUE
Or, Two Flies in the Ointment:
Communist China And Nazi Germany

∴

"The relationship between two assets can never be captured by a single scalar quantity."
—Financial Analyst Paul Wilmott[1]

A. Confronting the Beast

"Modern fiat money and reserve banking is indeed a manifestation of the transmutative 'nothingness' of the Philosophers' Stone, for from the creation of credit out of nothing, gold is produced." By nationalizing that money and credit-creating institution "and wresting it from private, secretive hands, and using it to fund the alchemical physics it was beginning to develop as the ultimate energy source, as the ultimate power to transport mankind, and as the ultimate power for destruction on a doomsday scale, the Nazis indicated that they had understood the nature of the (Philosophers') Stone. *They had seen, and fully understood, the connection between alchemical physics, and alchemical finance. And they were willing to put it to supremely evil uses.*

"But that connection between alchemical physics and alchemical finance is, perhaps, a relationship that requires its own exposition....

"*Epilogue is Prologue...*"

Thus did I write at the conclusion of my book *The Philosophers' Stone: Alchemy and the Secret Research for Exotic Matter.*[2] The reader may have inferred

---
1 Cited in Felix Salmon, "A Formula for Disaster," *Wired,* March 2009, p. 112.
2 Joseph P. Farrell, *The Philosophers' Stone: Alchemy and the Secret Research for Exotic Matter* (Feral House, 2009), p. 337, emphasis added.

from these quoted remarks that there was much more of the story — both from the standpoint of physics and finance, and from that of history — to tell, and that it would require yet another study or book to do so. If the reader made such inferences, he is correct on both counts: there *is* much more of the story of the relationship between physics and finance to tell.

The thesis of this book is both simple to state, and difficult to understand, and that is that, since ancient times and with more or less uninterrupted constancy, there has existed an international money power which seeks by a variety of means — including fraud, deception, assassination, and war — to usurp the money- and credit-creating power of the various states it has sought to dominate, and to obfuscate and occult the profound connection between that money-creating power and the deep "alchemical physics" that such power implies.

Accordingly, I do not argue that case comprehensively in this book, since to do so would require an extended series of books, each devoted to a particular historical period, and each burying the reader in a blizzard of footnotes to the extent that the main thesis would itself become obscured. Rather, I assume this model as a given, as an *interpretive paradigm* by which to view certain events and data. In so doing, that case is indeed argued, but in synoptic form rather than comprehensively. In doing so, I hope to keep before the reader's attention that deep and profound connection between physics and finance and to show why it is that the private and international money power must always seek to suppress not only certain types of state financial policy, but also certain types of physics, for both indeed spring from a common conceptual root.

Most of my books, as readers familiar with them already know, inhabit a strange region where alternative physics interfaces with history to reveal the latter's hidden motivations, secrets, and players. This book is no different, save for the fact that I have obviously added a new conceptual player: finance and economics. And along the way, we shall encounter other major conceptual scenery that readers of my books have encountered before: alchemy, astrology, astronomy, torsion, Egypt, Babylon, Nazis, ancient texts and tomes and modern mathematical gurus speaking the arcane language of statistical and topological lore.

In fact, in one of those odd synchronicities that seem to increase in modern life, as this book was being researched and written, decades — if not centuries or even millennia — of corruption and intellectual flaccidity in the financial, banking, and corporate sectors of the world came to an ugly head with the collapse of the housing and derivatives bubble, and the appearance of some of those responsible for the meltdown before the United States House of Representatives, hands extended, asking for a bailout of their malfeasance and irresponsibility at the expense of the American taxpayer, and demanding no oversight to boot, as if they were being forced to pay some hidden blackmailer, and were afraid that oversight might disclose this fact.

But why call it "irresponsibility" and not simply "criminality"? In the answer to that question there lies a tale, and it is a tale I did *not* originally intend to go into when I conceived the plan for this series of books many years ago, much less the plan for *this* one. Recent financial events, however, have contrived to place the story I intended to tell after completing *The Nazi International* and *The Philosophers' Stone* into a rather different context. As will become apparent to the reader in the main text, I *do* believe there is criminality and conspiracy involved in the story of the complex relationship of physics and finance throughout history. And paradoxically, the farther back one pursues this relationship, the *closer together* physics, finance, and all those other themes enunciated above as the conceptual scenery, draw together, and the more apparent the odor of a long-standing conspiracy becomes.

But in and of itself the contemporary financial meltdown is both a story of conspiracy as well as a case of galloping stupidity and colossal intellectual, political, and economic irresponsibility proportional to the aforesaid stupidity. It is nonetheless a story with its own deep connections to the story of the main text, and it is as good an entry into the subject as any.

So, as a way of entering into the discussion of the themes that preoccupy the main text, one may examine two salient modern examples that arose to challenge the reigning financial and physical assumptions of that money power.

Those examples are Communist China and Nazi Germany.

B. Meltdown, or Message
1. David Li's Formula

This most contemporary chapter of this very long and ancient story began when a Chinese mathematician, Dr. David X. Li, came up with a formula that seemed a godsend to Wall Street and City of London financial manipulators. The formula, known to most economists and mathematical analysts within major banking houses, is unfamiliar to most, but like Einstein's $E=Mc^2$ it is destined to become famous in history for the influence it had on human actions and politics. Here's what it looks like:

$$Pr[T_A<1, T_B<1] = \phi_2(\phi^{-1}(F_A(1)), \phi^{-1}(F_B(1), \gamma).$$ [3]

---

3 Felix Salmon, "A Formula for Disaster," *Wired,* March 2009, pp. 78–79.

Li, who "grew up in rural China in the 1960s"[4] eventually earned a master's degree in economics at Nankai University and then left China to pursue an MBA from Laval University in Quebec, a master's degree in actuarial science and a Ph.D. in statistics from the University of Waterloo in Ontario, Canada.[5] By 1997 Dr. Li had "landed at Canadian Imperial Bank of Commerce," later moving to Barclays Capital in 2004.[6]

But what did the formula actually mean? What did it *do?*

Felix Salmon summarizes the effect of the formula in a brilliant article, "A Formula for Disaster," in the March 2009 issue of *Wired* magazine.

> (Li) took a notoriously tough nut — determining correlation, or how seemingly disparate events are related — and cracked it wide open with a simple and elegant mathematical formula, one that would become ubiquitous in finance worldwide.
>
> For five years, Li's formula, known as a Gaussian copula function, looked like an unambiguously positive breakthrough, a piece of financial technology that allowed hugely complex risks to be modeled with more ease and accuracy than ever before. With his brilliant spark of financial legerdemain, Li made it possible for traders to sell vast quantities of new securities, expanding financial markets to unimaginable levels.
>
> His method was adopted by everybody from bond investors and Wall Street banks to ratings agencies and regulators. And it became so deeply entrenched — and was making people so much money — that warnings about its limitations were largely ignored.
>
> Then the model fell apart. Cracks started appearing early on, when financial markets began behaving in ways that users of Li's formula hadn't expected. The cracks became full-fledged canyons in 2008 — when ruptures in the financial system's foundation swallowed up trillions of dollars and put the survival of the global banking system in serious peril.
>
> ....Li's Gaussian copula formula will go down in history as instrumental in causing the unfathomable losses that brought the world financial system to its knees.[7]

But again, what precisely did the formula *do?* And *how* did it do it? How did it cause the financial meltdown?

---

4 Ibid., p. 78.
5 Ibid.
6 Ibid. While Salmon does not mention it, it should be noted that the Canadian Imperial Bank of Commerce has ties to the Rothschild banking interests, some of whose directors have later worked for the CIBC.
7 Felix Salmon, "A Formula for Disaster," *Wired,* March 2009.

The key lies precisely in that important word, "correlation." Li's formula seemed to make sense of the "thousands of moving parts"[8] of an interlocked economic system. Salmon illustrates what Li's formula did via a simple analogy:

> To understand the mathematics of correlation better, consider something simple, like a kid in an elementary school. Let's call her Alice. The probability that her parents will get divorced this year is about 5 percent, the risk of her getting head lice is about 5 percent, the chance of her seeing a teacher slip on a banana peel is about 5 percent, and the likelihood of her winning the class spelling bee is about 5 percent. If investors were trading securities based on the chances of those things happening only to Alice, they would all trade at more or less the same price.
>
> But something important happens when we start looking at two kids rather than one — not just Alice but also the girl she sits next to, Britney. If Britney's parents get divorced, what are the chances that Alice's parents will get divorced, too? Still about 5 percent: The correlation there is close to zero. But if Britney gets head lice, the chance that Alice will get head lice is much higher, about 50 percent — which means the correlation is probably up in the 0.5 range. If Britney sees a teacher slip on a banana peel, what is the chance Alice will see it, too? Very high indeed, since they sit next to each other: It could be as much as 95 percent, which means the correlation is close to 1. And if Britney wins the class spelling bee, the chance of Alice winning it is zero, which means the correlation is negative: -1.
>
> If investors were trading securities based on the chances of these things happening to both Alice *and* Britney, the prices would be all over the place, because the correlations vary so much.[9]

Now factor in thousands, even millions, of individuals, and thousands of interlocking correlated conditions — energy prices, building and housing costs, money and credit supply and so on — and one gets the idea of the complexity of the system of correlations, and the beguiling simplicity of Dr. Li's formula.

The reason its simplicity was so beguiling is in itself rather simple. As Salmon puts it in his article:

> …(It's) a very inexact science. Just measuring those initial 5 percent probabilities involves collecting lots of disparate data points and

---

8 Ibid., p. 77.
9 Ibid., pp. 77–78.

subjecting them to all manner of statistical and error analysis. Trying to assess the conditional probabilities — the chance that Alice will get head lice *if* Britney gets head lice — is an order of magnitude harder, since those data points are much rarer. *As a result of the scarcity of historical data, the errors there are likely to be much greater.*[10]

Bear that point about the need for historical data in mind, for it not only plays a significant role immediately below in the methodological assumptions of Dr. Li and his formula, but will play an even more important role in chapter two.

It is regarding this point of historical data, or rather, the *assumed* lack thereof, that Dr. Li's formula provided a way out of the impasse, or so it seemed at the time:

> Using some relatively simple math — by Wall Street standards, anyway — Li came up with an ingenious way to model default correlation *without even looking at the historical default data.* Instead, *he used market data about the prices of instruments* known as credit default swaps.[11]

In other words, within Li's elegant statistical copula formula lies a hidden methodological assumption, namely, that historical data on credit default rates could be safely jettisoned in favor of an immediate concentration on the "current" market prices of "credit default swaps." As we shall see in chapter two, there was a great deal of historical data available, and that data in turn pointed to a "deep physics" of financial cycles that few economists — or even physicists for that matter — could scarcely guess at and even fewer knew existed at all.

But we are getting ahead of ourselves. What exactly are "credit default swaps"? This is where the story of Li's formula gets very interesting, and Salmon explains what they are with clear concision:

> If you're an investor, you have a choice these days: You can either lend directly to borrowers or sell investors credit default swaps, *insurance against those same borrowers defaulting.* Either way, you get a regular income stream — interest payments or insurance payments — and either way, if the borrower defaults, you lose a lot of money. The returns on both strategies are nearly identical, *but because an unlimited number of credit default swaps can be sold against each borrower, the supply of swaps isn't constrained the way the supply of bonds is, so the*

---

10   Felix Salmon, "A Formula for Disaster," *Wired,* March 2009, 74–79, 112, p. 78, emphasis added.
11   Ibid., emphasis added.

CDS market managed to grow extremely rapidly. Though credit default swaps were relatively new when Li's paper came out, they soon became a bigger and more liquid market than the bonds on which they were based.[12]

Enter Dr. Li, for his formula was nothing more than "a model that used price rather than real-world default data as a shortcut (making an implicit assumption that financial markets in general, and CDS markets in particular, can price default risk correctly)."[13] Salmon explains Li's technique by mincing no words:

> It was a brilliant simplification of an intractable problem. And Li didn't just radically dumb down the difficulty of working out correlations; he decided not to even bother trying to map and calculate all the nearly infinite relationships between the various loans that made up a pool. What happens when the number of pool members increases or when you mix negative correlations with positive ones? Never mind all that, he said. The only thing that matters is the final correlation number — one clean, simple, all-sufficient figure that sums up everything.[14]

Li had reduced the problem to a simple dimensionless number — a scalar in mathematicians' terms — that in its simplicity cast a beguiling spell over the world's financial and securities market.

Indeed, the formula's effect was almost immediate and "electric" because Wall Street's financial gurus, "armed with Li's formula" deduced from it a

> ...new world of possibilities. And the first thing they did was start creating brand-new triple-A securities. Using Li's copula approach meant that ratings agencies like Moody's — or anybody wanting to model the risk of a (bundle of securities) — no longer needed to puzzle over the underlying securities. All they needed was the correlation number, and out would come a rating telling them how safe or risky the (bundle) was.
>
> As a result, just about anything could be bundled and turned into a triple-A bond — corporate bonds, bank loans, mortgage-backed securities, whatever you liked. The consequent pools were often known as collateralized debt obligations, or CDOs. You could (bundle) that pool and create a triple-A security *even if none of the*

---

12 Felix Salmon, "A Formula for Disaster," *Wired,* March 2009, 74–79, 112, p. 78, emphasis added.
13 Ibid.
14 Ibid., pp. 78–79.

> *components were themselves triple-A.* You could even take lower-rated (bundles) of **other** CDOs, put them in a pool, and (bundle) them — an instrument known as a CDO-squared, which at that point *was so far removed from any actual underlying bond or loan or mortgage that no one really had a clue what it included.* But it didn't matter. All you needed was Li's copula function.[15]

In other words, the effect of Dr. Li's formula was to abandon the focus on the relative strength and risk of each component, since they were now interlocked via the copula itself, and this in turn led to an explosion of more and more "bundles" of securities and credit swaps, and even to bundles of bundles.

As a result of this increasing interlock and correlation,

> The CDS and CDO markets grew together, *feeding on each other.* At the end of 2001, there was $920 billion in credit default swaps outstanding. By the end of 2007, that number had skyrocketed to more than $62 **trillion.** The CDO market, which stood at $275 billion in 2000, grew to $4.7 trillion by 2006. At the heart of it all was Li's formula.[16]

To put it succinctly, Li's formula was a way of adding yet another multi-layered ability to create credit and interest — an alchemical operation — out of literally nothing.

There was also another hidden danger in Li's formula, and that was that even for those financial institutions that *did* take into consideration historical data, there was not much historical data to go on for the simple reason that "credit default swaps had been in existence less than a decade," a decade during which "house prices soared."[17] Indeed, Li's formula had the effect, as already has been seen, of inflating such "bundling" exponentially and contributing to the creation of bundles of bundles with literally millions of potential correlated factors, *none of which*, it bears repeating, were based on historical data of anything more than a few years' duration. And this reveals another flaw in the formula and its application to securities correlations and their ratings, for implicit in the technique was a hidden assumption that "correlation was more of a constant than a variable,"[18] that is, while the individual components of such bundles *were* variable, their correlation was *not.*

---

15 Felix Salmon, "A Formula for Disaster," *Wired,* March 2009, 74–79, 112, p. 79, italicized emphasis added, bold emphasis in the original.

16 Felix Salmon, "A Formula for Disaster," *Wired,* March 2009, 74–79, 112, p. 79, italicized emphasis added, bold emphasis in the original.

17 Ibid., p. 112.

18 Ibid.

Consequently, by abandoning historical data, and by relying upon this hidden assumption of correlation constancy, the model *could not* cope with any sudden downturn in prices, such as occurred in the housing market and mortgage sectors. In short, the whole technique had abandoned the well- and commonly-known fact that economic activity, for whatever reason, seems to occur in repeated *cycles* of growth and decline, or if one prefer, of "boom" and "bust." And in this knowledge of cycles, as will be seen, there also lies quite a tale, and a carefully hidden one at that.

Suspiciously, Dr. Li, after publishing his formula, returned to China in 2008 and has been curiously silent during the debate over the causes and culprits behind the crash. But in the ultimate twist to the story, he has returned to Beijing where he is in charge of "the risk-management department of China International Capital Corporation"![19] This raises disturbing possibilities, not the least of which was that the whole episode might conceivably have been a form of economic warfare, but by whom, and against whom?

Whatever the answer to that question may be — and it is not as apparent as it might seem — it is clear that Dr. Li's formula would provide the necessary mathematical technique and "technology" to anyone inclined to wage such economic warfare, particularly if such persons or groups were in the possession of historical or other data that indicated an inevitable economic downturn were coming, and who, seeking to worsen it for their own purposes, used Li's formula to create the correlatives or derivatives bubble of bundles and credit swaps, a bubble that would pop when the inevitable fall in prices occurred. As will be seen in chapter two, there is precisely such data available, and it did indicate an inevitable cyclic economic downturn would begin ca. 2000–2006.

2. *The Li Clan, the Canadian Imperial Bank of Commerce, and the Triads*

But that is not all there is to the David Li story. This part of the story, however, appears to have been carefully hidden from the public, for reasons that will become apparent in a moment. As noted above, among Dr. Li's many career moves and positions, he began to work for Canadian Imperial Bank of Commerce in 1997, moving from there to Barclay's in 2004. Thus, when his paper with the notorious formula — "On Default Correlation: A Copula Approach" — was published in 2000, Li still worked for Canadian Imperial Bank of Commerce.

This reveals some interesting possible connections.

As most people know, Chinese names actually begin with the surname,

---

19 Felix Salmon, "A Formula for Disaster," *Wired,* March 2009, 74–79, 112, p. 112.

followed by the individual's proper name. Thus my name, following the Chinese custom, would be Farrell Joseph rather than Joseph Farrell. More importantly, the name Li is a fairly common surname in China, like the name Smith or Brown would be in the U.S.A., Britain, or Canada. People with a common surname do not therefore imply any blood relation, and people with the same surname are often strangers to each other.

The same holds true for China, but with one significant exception. In Chinese culture, people with the same surname — regardless of whether there are blood ties or not — are regarded by the Chinese as coming from the same family, or clan. Thus, two Chinese strangers with the same surname will more often than not regard each other differently than two strangers with *different* surnames. They will, to a certain extent, regard each other as part of a very large family, extending to each other the customs and courtesies common between family members.

And this places Dr. David X. Li into a very different potential interpretive context, for at the same time as he was employed by the Canadian Imperial Bank of Commerce and authoring his now notorious paper, a fellow clansman, one Li Kai-Shing, owned a significant bloc of that bank's stock.

But who is Li Kai-Shing? He is one of Hong Kong's most famous billionaires, who, along with his sons Victor and Richard, assumed significant governmental posts to aid in the transition of Hong Kong from a British Crown Colony back to Chinese jurisdiction. Li Kai-Shing's financial empire stretches through Asia's media and financial markets, and even more significantly, another Li, Li Chiang, was at one time head of Red China's China International Trust and Investment Corporation, in charge basically of China's foreign trade with the West, and particularly with the United States and Canada.[20] After Dr. David Li left the Canadian Imperial Bank of Commerce in 2004, Li Kai-Shing sold his stake in the bank the very next year.[21]

Not even this, however, begins to plumb the depths of the importance of the Li clan in China, for the clan has had several emperors, including Emperor Li Zhuanxu, who reigned before 2000 B.C. Additionally, a Li founded the Tang Dynasty (618–906 A.D.). Most significantly, the Li clan appears to have been involved with financial wheeling and dealing from an early time, being the same clan that introduced paper money to China during the same Tang dynasty.[22]

We now have the following intriguing latticework of relationships:

---

20   Q.v. Fritz Springmeier, *Bloodlines of the Illuminati* (Ambassador House, 2002), pp. 163–185.
21   www.absoluteastronomy.com/topics/Canadian_Imperial_Bank_of_Commerce. This website also clearly implicates the CIBC in the Enron scandal.
22   Springmeier, op. cit., pp. 164–165.

1) An old and influential Chinese clan, very much involved in government and finance, from early Chinese history;
2) The same clan involved in the same activities millennia later; and
3) One member of the clan develops *the* formula that led to the current meltdown. And far from being *dismissed* as a competent risk assessor by the Chinese, he returns to China to a post doing precisely what his formula was designed to do: assess risk!

All this, of course, places David X. Li's return to Red China and assumption of a position in a Chinese corporation concerned with overseas trade, and once again in a position where he is "assessing risk," into a very interesting context, making it unlikely that his and namesake Li Kai-Shing's departure from the Canadian Imperial Bank of Commerce were coincidental.

But not even this tells the whole tale of the Li clan!

The Li clan "is one of the principal families which has controlled" the notorious Chinese criminal secret society, the Triads.[23] According to researcher Fritz Springmeier, the following Lis have been leaders of various sections of the Triads:

Li Chi-t'ang	-overseas leader
Li Hsien-chih	
Li Hsiu-ch'eng	-Hunan
Li Hung	-Honan
Li K'ai-ch'en	-Triad, Shanghai
Li Lap Ting	-Kwangsi province
Li Ping-ch'ing	-Triad, Shanghai
Li Shih-chin	
Li Wen-mao	-north of Peking, Fatshan
Li Yuan-fa	-Hunan
Li Choi-fat	-Hong Kong
Li Jahfar-Mah	-Britain[24]

And just what are the Triads? They are a Chinese criminal secret society, deeply connected with the opium trade, the practice of Chinese occult activities, and they are, in Springmeier's apt description, "something of a cross between the Masons and the Mafia — something in the line of P2 Freemasonry — except much bigger."[25] Indeed, some estimates put the membership of the Triads and

---

23   Ibid., p. 176.
24   Springmeier, op. cit., p. 176.
25   Ibid. P2 Masonry refers to the notorious Lodge Propaganda Due, led by Mason Licio Gelli,

similar Oriental secret societies as close to two million, if not more. Thus, the possibility that the entire current meltdown was the a covert act of economic warfare looms rather large, especially given the fact that the Canadian Imperial Bank of Commerce is also known to have some strong ties to the Rothschild banking interest.

In short, we have clear indications of the possibility that China has signaled that it is not simply going to be a subservient player in whatever "New World Order" schemes the Anglo-American elite wishes to implement and impose on the rest of the world.

In this respect, it is important to observe the pattern that the Li clan typifies:

1) The clan is ancient, with strong ties going back millennia, both to government and finance;
2) The clan is moreover connected with a secret society, whose activities in turn are connected with:
   a) Occult religious activity;
   b) Criminal business organization and activity;
   c) Assassinations, blackmail, and infiltration of government and finance;
3) The clan is clearly connected to a mathematical model of economic and credit activity, implying a hidden interest in developing such formally explicit models; and finally,
4) That model is *the* culprit responsible for the current economic meltdown, which has affected primarily the institutions of the Anglo-American financial elite.

As will be seen, such activities and patterns or relationships are not unique to the present or previous century. They are, in fact (as the Li clan itself evidences) millennia old and firmly rooted within the history of human civilizations and their banking class. However, the consistency of this behavior and this constellation of relationships is rooted in "something more" — and certainly nothing less — than "human nature," or "greed," or even "the predictable actions of a particular socio-economic class."

---

that was finally exposed by the Italian government in the 1980s. By the time the Italian authorities had shut the lodge down, it numbered over two thousand members, with deep penetration into banking, the Italian government, various Italian political parties, and the Vatican. Many who regard the death of Pope John Paul I as suspicious think that the Pope was murdered precisely because he discovered the extent of Masonic P2 penetration of the Vatican and was determined to root it out and end it. (Q.v. David Yallop's *In God's Name.*) Many also suspect that the strange death of banker Roberto Calvi, hung by the neck from London's Blackfriars' Bridge, and in self-evidently Masonic fashion, was also tied to the P2 scandal. Also prominent in the Lodge were various Italian and other Fascists. Not surprisingly, Licio Gelli, its founder, fled to Argentina after the lodge was exposed.

That something else lies in a little-noticed connection between physics and finance that is as old as man's fascination with the stars, and with what good or ill fortunes they might portend. And that is also to say that there has been, since ancient times, a profound connection between physics and economics, and a struggle between those who view both as closed systems, and those who view both as open systems. We shall call those who adhere to the closed system of physics and finance by the term "banksters," to indicate the chimerical and criminal hybrids of gangster and banker. These people are indeed driven by an almost boundless lust for power, by a criminal greed and wanton disregard of the humanity they wish to enslave. But they are also driven by a knowledge of certain hidden things, among them the profound relationship between physics and finance.

3. China's Money

There is a little-known aspect to China's booming economy that Western financiers, economists, and media mandarins are loath to discuss, and that is that China's money is created by *China*, and *not* borrowed from private bankers. In effect, this means China's money is *debt-free*. In short, Communist China has followed the precedent established by the American federal constitution, where the money-creating and issuing power lies with the Congress, that is, with the state. As Ellen Hodgson Brown observed in her magisterial book on the whole problem of central private banking monopolies on money issuance, *Web of Debt,* the key difference in China's system and that prevailing in most other nations

> Is its banking system. China has a government-issued currency and a system of national banks that are actually owned by the nation. According to Wikipedia, the People's Bank of China is "unusual in acting as a national bank, *focusing on the country not on the currency."* The notion of "national banking," as opposed to private "central banking," goes back to Lincoln, Carey, and the American nationalists. Henry C.K. Liu distinguishes the two systems like this: a national bank serves the interests of the nation and its people. A central bank serves the interests of private international finance.[26]

Even though China's currency, the yuan, is pegged to the dollar in terms of its exchange rate value, China's banking system is testament to the fact that its

---

26 Ellen Hodgson Brown, *Web of Debt: The Shocking Truth about Our Money System and How We Can Break Free* (Baton Rouge, Louisiana: 2008), pp. 254–255.

government has seen what American politicians and a great deal of its people *formerly* understood, namely that when a private bank creates money, it *only creates the principal, not the interest.* When a state, on the other hand, creates money, it has the ability to expand the money supply according to the credit needs of the nation.

In short, when a nation's money represents *a principal on which interest is owed, someone always comes out the loser, since there is never enough money in circulation to repay the debt interest, and thus, a national debt can never be repaid, it can only grow.* Contrariwise, when a nation's money represents *a receipt for goods and services rendered and is issued interest-free by the state itself, that state can experience almost total employment, and there is no built-in principal of debt and scarcity.*

It is these two systems, where money is created by a private monopoly in limited supply, which represents an interest-bearing debt note, or where money is created by the state as a receipt for goods and services and is *debt-free,* that are the two systems which have contended against each other throughout history. In the former system, the system of money in circulation is *closed,* and there is never as much money in circulation as there is debt, and hence, scarcity becomes the order of the day, as limited supplies of money compete for limited goods, resources, and energy. In the latter system, the system of money is *open* and can expand, as the economy whose goods and services it represents as receipts expands with it. In terms of the analogy to physics, the former system can *never* function at over-unity, whereas in the latter system, it *must* function as such.

C. Nazi Germany: Physics and Finance Fully Rationalized

This relationship between finance and physics was, in modern times, first clearly perceived by that nation which not only established state-created debt-free money, but which also sponsored a variety of secret research projects into "free-energy" physics and technologies: Nazi Germany.

When World War I ended and the Allies imposed war reparations on defeated Germany, the total reparations to be paid exceeded the value of all the property in Germany by three times![27] Anyone who has studied the history knows the story: Germany hyper-inflated its currency, paying off the Allies with increasingly worthless Reichsmarks and destroying Germany's economy in the process. But, it is to be noted, these Reichsmarks were still the issuances of a privately controlled bank, and thus, Germany's debt situation only compounded dramatically.

---

27 Hodgson, *Web of Debt*, p. 229.

Hitler's Reichsbank President, Dr. Hjalmar Horace Greeley Schacht, let the big secret out in his 1967 book *The Magic of Money:*

> The mark's dramatic devaluation began soon after the Reichsbank was "privatized," or delivered to private investors. *What drove the wartime inflation into hyperinflation, said Schacht, was speculation by foreign investors, who would sell the mark short, betting on its decreasing value....* Speculation in the German mark was made possible because the Reichsbank made massive amounts of currency available for borrowing, marks that were created on demand and lent at a profitable interest to the bank. When the Reichsbank could not keep up with the voracious demand for marks, other private investment banks were allowed to create them out of nothing and lend them at interest as well.[28]

Thus, the German government was not responsible for the postwar hyperinflation. It was Germany's *privately owned central bank and its monopoly on the country's money creation — money created as a circulating note of debt — that created the problem!* Germany's economy was crashed and devastated by the bankers.

Until Hitler.

While most people are aware that various private financial powers in the West were instrumental in placing Hitler and the National Socialist Party into power in Germany,[29] what most do not realize is how quickly Hitler turned on his backers and refused to play ball by the same old Rockefeller-Rothschild rules:

> ...(A)utocratic authority did give Adolf Hitler something the American Greenbackers could only dream about — total control of the economy. He was able to test their theories, and he proved that they worked. Like for Lincoln, Hitler's choices were to either submit to total debt slavery or create his own fiat money; and like Lincoln, he chose the fiat solution. He implemented a plan of public works along the lines proposed by Jacob Coxey and the Greenbackers in the 1890s. Projects earmarked for funding included flood control, repair of public buildings and private residences, and construction of new buildings, roads, bridges, canals, and port facilities. The projected cost of the various programs was fixed at one billion units of the national currency. One billion non-inflationary bills of exchange called Labor

---

28 Hodgson, *Web of Debt*, p. 233.
29 See, for example, the seminal work and research of Anthony Sutton, *Wall Street and the Rise of Hitler.*

> Treasury Certificates were then issued against this cost. Millions of people were put to work on these projects, and the workers were paid with the Treasury Certificates. The workers then spent the certificates on goods and services, creating more jobs for more people. The certificates were also referred to as MEFO bills, or sometimes as "Feder money." ...(T)hey avoided the need to borrow from international lenders or to pay off international debt.

The result of these Nazi machinations against the international money power was predictable: foreign credit was refused, and hence, Germany faced an almost complete inability to conduct foreign trade and commerce. But again, the Nazi regime did an end run around the banksters, restoring foreign trade by cutting out the banking middleman and resorting to a system of barter with other nations.[30]

And where did Hitler get these "radical" financial ideas?

When he first attended a meeting of the early National Socialist Party, he learned of the views of a German economist named Gottfried Feder.

> The basis of Feder's ideas was that the state should create and control its money supply through a nationalized central bank rather than have it created by privately owned banks, to whom interest would have to be paid. From this view derived the conclusion that finance had enslaved the population by usurping the nation's control of money.[31]

Feder and other German theorists had for their part based their theories on a study of the American constitution, and more importantly, that President Abraham Lincoln had financed the Northern effort in the American Civil War by creating debt-free "greenbacks," bypassing New York banks and interest debt completely.[32]

But Nazi Germany did something else, something quite significant. Realizing that Germany was at the mercy of the very banksters that controlled the world's oil supplies and hence the energy needed to maintain Germany's national sovereignty, The Third Reich established entire departments of the SS called the *Forschung, Entwicklung, und Patente,* and the *SS Entwicklungstelle 4,* or "Research, Development, and Patents," and "SS Development Area Four," respectively. The top secret mission brief of these departments were to scour and pull patents having national security implications, and *to investigate and*

---

30 Hodgson, *Web of Debt,* p. 230.
31 S. Zarlenga, *The Lost Science of Money* (Valatie, New York: 2002), p. 590, cited in Hodgson, *Web of Debt,* p. 231.
32 Ibid.

develop the technologies of "free energy," i.e., the technologies that would allow Germany to engineer the physical medium and its energy directly, and to tap into it for its energy needs, and as a weapon.[33] Add to this the fact that, within the intellectual cauldron that was the SS, ideas from advanced though little-known physics conceptions circulated and percolated along with the idea that there was a physics hidden in ancient tomes and epics,[34] and one now obtains a very revealing set of relationships:

1) A nation that has restored its sovereign right to issue its own debt-free currency, breaking from the orbit of the international money power;
2) A nation that has also clearly seen that in order to break completely from the dominance of that power, it must have access to a completely different source of energy that is nearly inexhaustible and not monopolized by that private money power;
3) Thus, that nation must seek to develop and control the technologies of the manipulation of that energy for itself; and finally,
4) That nation has perceived *an ancient link* or connection between the physics it seeks to develop, and the financial policy it seeks to develop.

One may discern from this list quite obviously that Nazi Germany was not only able to achieve nearly full employment mere years after the regime took power, but that it was very deliberately gearing up for an inevitable war. But this list discloses a possible hidden *reason* for those war preparations beyond those of the Nazi lust of "living space" and world conquest; Germany's decision to issue state-created money "would mean that the international financiers would be unable to exercise...control through the international gold standard...and this may have led to controlling Germany through warfare instead."[35] This hidden reason for the war — that the Allies essentially acted as "agents" for the international money power against a great power that had essentially severed all connections to it — may also provide a similar rationalization for the Allied demands for German unconditional surrender, essentially ensuring that the war would continue until Germany was basically destroyed and physically occupied.

The relationship between banking and the quest to acquire, or suppress,

---

33 This story I have detailed in my previous books *Reich of the Black Sun, The SS Brotherhood of the Bell, Secrets of the Unified Field* (Adventures Unlimited Press), and *The Philosophers' Stone* (Feral House).
34 For this point, see my *Reich of the Black Sun*, pp. 161–180, and *The Philosophers' Stone*, part four.
35 S. Zarlenga, *The Lost Science of Money* (Valatie, New York: 2002), p. 590, cited in Hodgson, *Web of Debt*, p. 231.

exotic physics technology may be glimpsed by yet another detour, this time to a persistent struggle.

D. Energy and Money-Creating Autarchy

If one takes these two very disparate instances as a clue, and especially that of Nazi Germany, then this suggests that there has been a persistent struggle throughout history between those who wish to democratize the production of energy, based on "alternative energy technologies," and who, similarly, wish to restore to the state its money-creating power and wrest it from private hands, and those who wish to monopolize that hidden technology and physics, and similarly, to hold a private monopoly over the money-creation of various states.

Consider what these very unique and different examples disclose:

1) In the case of Communist China, we have an instance of a modern and technologically sophisticated world power issuing state-created debt-free money, a fact that has led to its economic boom and its "independence" from the international money power;
2) In the case of Nazi Germany, we have not only a nation that saw the advantage of such state-issued debt-free currency, but also of a state that clearly saw the connection between that sovereign money power and the analogous physics of receiving its energy *directly* from the physical medium, an energy source *not* based on non-renewable energy "resources." Furthermore, it is clearly implied by the Nazi SS' interest in ancient and esoteric texts that there may be an ancient connection between this type of finance and this type of physics;

As will be seen, the struggle between these two camps has erupted throughout history in violence, as the latter group that advocates open systems seeks to overturn the dominant order of the moneychangers — the banksters — based on closed systems, or, conversely, as the banksters seek to extend their power via closed systems of physics and economics must respond to the inevitable threat posed by civilizations or countries adopting the open ones. Most recently that struggle erupted in the enormous conflict we call World War II, as Nazi Germany, for whatever its genocidal and murderous crimes against humanity, at least perceived *part* of the struggle correctly: it was a war to free Germany from a heinous international money power — misrepresented of course in Nazi ideology by the Jews — based in Great Britain and the British Empire, and in the United States of America.

Thus, Nazi Germany's pursuit of "free energy" and "energy independence" or "autarchy," as well as her pursuit of a radical alternative hyper-dimensional physics, was a part of that struggle.[36] By the same token, not for nothing did Nazi Germany essentially restore the idea of public, debt-free, *state-created* money and credit, while simultaneously pursuing these super-secret advanced physics projects. The two were conceptually united, and, as we shall also discover within these pages, had been so for a long time. The Nazis were simply the ones who in modern times first drew the connection between physics and finance, and determined to do something about it. As far as Nazi Germany was concerned, then, there was a grisly and gruesome logic to the Holocaust, for having identified "World Jewry" as the center of this International Money Power, the use of enslaved Jews in the concentration camps to create this new physics and its associated technologies, and a new economy on which they hoped eventually to place the Reich, was in their warped view a kind of "justice" meted out to those who had so ravaged and pillaged Germany in the wake of the Versailles Treaty and the Dawes and Young Plans. There was, of course, a half-truth to this, for there was indeed a prominent element of Jewish, or rather, Zionist influence in such financial circles. One need only recall the names of Rothschild, Warburg, Schiff and so on in this regard. But there all truth stops and the lies begin, for there was also a prominent element of Protestant monied groups involved as well. Thus, for consistency's sake, the Nazis should equally have been concerned with the Protestant aristocracy of England, America, Canada, and even their own country as well.

World War II ended, of course, with the defeat of Nazi Germany, but not necessarily with the defeat of *Nazism* and its goals in alternative physics and finance.[37] However, as we shall see, the postwar period began with an unusual event, an event I believe many to have profoundly misinterpreted. Rather than seeing that a kind of *détente* or *modus vivendi* was struck between elements of what I call the "Nazi International" on the one hand and the globalist Anglo-American corporate and banking elite on the other, most prefer to view certain events as testament to the fact that one more or less coherent and monolithic "international conspiracy of money" exists, and that it is entering the "end game phase" of its goal of total world domination. The closer to that goal that this group actually approaches, the more sharp and acute will factional infighting within it become, particularly as such banksters are in thrall to the closed system paradigm of physics and finance. In short, that postwar détente is showing

---

36 For the stories of this pursuit, and the bureaucracies they emplaced to pursue it, see my *The SS Brotherhood of the Bell*, Adventures Unlimited Press (2005); *Secrets of the Unified Field*, Adventures Unlimited Press (2007); *The Nazi International*, Adventures Unlimited Press (2008); and *The Philosophers' Stone*, Feral House (2009).

37 See my *The Nazi International: The Nazis' Postwar Plan to Control Finance, Conflict, Physics, and Space*, Adventures Unlimited Press (2008), pp. 249-350.

distinctive signs of breaking down, as each faction maneuvers to emerge as the dominant faction when the eventual goal is reached. For that globalist elite, this problem is further compounded by the emergence of the economic powers of Eurasia — Russia, China, and Japan — who are also showing increasing signs of reluctance to play by the tired old rules of the Anglo-American empire. Witness once again Dr. Li's mysterious "disappearance" from the West and reemergence as a risk manager for a prominent corporation in Communist China!

However, between those periods when this conflict of worldviews of physics and finance breaks out into open violence and warfare, the struggle is more covert and hidden, as suggested, again, by the David Li episode. In the case of such covert warfare, the various factions of international banksters resort to every occulted means at their disposal to suppress open development of alternative physics — and therefore of the alternative economies and financial institutions such physics would inevitably usher in — while they seek to maintain the status quo and become the dominant faction within it. The methods, chiefly, are threefold: they must first suppress *technologies* that testify to the existence of another physics than the "public consumption physics" they have so carefully inculcated for the masses and promoted in academies and textbooks. Secondly, they must suppress *the alternative physics itself*, for in some respects it is the real source of their power, as we shall also eventually see. Thirdly, and finally, they must obfuscate the profound, deep, and ancient *connections* between this alternative physics and alternative institutions of finance and economy.

To sum it up, they must seek to suppress its open development, while simultaneously pursue its covert development so that they can, in turn, *monopolize* it and further consolidate power into their own hands. By the same token, this means that any given faction within what is called "The New World Order" must likewise inhibit or arrest the development of such alternative theories and technologies by *rival* factions, or alternatively, develop superior versions of it, or defenses against it, themselves. Conversely, those groups or individuals advocating open systems of physics and economics not only must run the gauntlet of all who oppose them, but develop their theories and technologies and bring them to as wide a public as quickly as possible. In short, the "good guys" seek to democratize the whole science.

Consequently, we are conspicuously in the presence of a very complex dynamic, one stretching from the individual person to whole civilizations and all the institutions thereof, and one moreover crossing several disciplines — physics, economics and finance, theology, history — and running like a gold and crimson thread throughout the millennia of human history. We must consider salient modern events where this struggle has openly erupted, and compare them to ancient manifestations of a similar nature. We must ponder the presence within ancient megaliths and temples of a profound astronomical and astrological science, their physics implications, and ponder further *why so many of these ancient sites are also associated with the presence of moneychangers, of banksters*. We must similarly ponder again what all this might mean for the occurrence of an ancient and interplanetary war within our own celestial neighborhood.[38] Yet again, we must consider what all this has to do with the stubborn persistence of alchemy from ancient to modern times, and what it may have to do with the consistent royal and imperial patronage of it throughout the Middle Ages and early Renaissance. We must look for the clues of deliberate suppression of this physics — and its implied economics — throughout the ages, and especially in our own. And finally, we shall have reason to consider why bloodlines seem to be such an important part of the story for those bankster and royal elites.

Given all these complex dynamics and disparate facts, I was confronted with something of a problem in writing this book. Normally I aim for a relatively high degree of "completeness" or at least thoroughness when writing a book. But I quickly discovered that, if I were to explore every facet of this complex dynamic with anything approaching thoroughness, I would not be writing a mere book, but *several* books, with each aspect of the problem requiring volumes of its own. For example, the involvement of major corporations in suppression of inventions and technologies implying a new physics and new energy source is a story that would and could consume volumes, as there is no shortage of such stories on the internet and books about the subject. Similarly, the machinations of the international banksters has provided a rich field for research and speculation, and spawned literally hundreds of books on the subject, both those by "sanctioned insiders" and those examining the subject from without. When one adds economics and physics and systems of money creation to this picture, the bibliography of such a work itself would end up being a whole book.

So clearly some other approach was needed. What I have attempted to do, therefore, is *outline a case and my interpretation of the evidence* backing it up with examples. This book is thus deliberately intended to be read not only in conjunction with my own previous books on alternative science and history as yet another chapter in a very big story, but also in tandem with the overall output of other researchers into the field. Thus, it simply *assumes* the existence of that research, and the reader's broad familiarity with it.

But let there be no mistake: this is not an easy, lighthearted book simply because it is not a *complete or thorough* one. The number and types of details, conceptions, and disciplines to be examined are considerable, and their

---

38 See my *The Cosmic War: Interplanetary Warfare, Modern Physics, and Ancient Texts*, Adventures Unlimited Press (2006).

interrelationships are even more so. With that said, the reader is cautioned about two things: first, in order to survey these connections and their implications, much of this book is "introduction"; only at the very end and in the final chapter will it be possible to tie all the threads together and draw their implications. Thus, patience is required as the data is laid out, and connections are drawn. Secondly, what is presented here is likewise a speculative though nevertheless argued case. Were each and every point to be documented at length, as already mentioned, each would require a tome in its own right.

This is a survey, not an encyclopedia; a study, not a painting; an essay, not a mathematical or historical proof.

Nonetheless, I hope that by consulting certain sources, mentioning certain topics, construing the interrelationships, and drawing the conclusions and implications in the manner that I have, that I will point the reader in the direction to examine these questions more fully on his own, for the problem is not in the *dearth* of information, interpretations, and implications, but in their *surfeit*. The interpretation proffered here is consequently not the only possible one. I maintain only that, of all the possible interpretations of this vast complex of information — alchemy, astrology, astronomy, torsion, Egypt, Babylon, Nazis, finance, geometries, earth grids and "scalar" physics, ancient texts and tomes and modern mathematical gurus speaking the arcane language of statistical economics and topological lore — that this interpretation is at least a plausible one.

All that being said, if along the way I have in some small measure contributed to the demise of these goofy, insane, and ruthless banksters and their murderously utopian and loony New World Order schemes, then so much the better, for one thing, I hope, is now evident: if in the private creation of money as an interest-bearing debt-note, only the principal, and not the interest, is created by those banks and circulated as "money," then it inevitably follows that, under such a system, debt can only *grow* and never be repaid. With that fact, the influence and control of that private class of banksters over the policies of a state can only grow, and to the increasing detriment of that nation's people and its public good.

As such, *any* discussion of financial policy by *any* politician of *any* political party affiliation in *any* nation that does not *begin* with a call to restore the power of money creation and issuance to the government in such a capacity that it is made to serve the public good and not the private rapacity of a bankster class, is merely deception, and deception for a very simple reason. The big secret of money is not only that it *must* represent some*thing*, but rather that it must represent some*one*. As such, there are really only two basic models of money known to history, the first model, where money represents a debt interest-bearing note — the "something" — created by a private monopoly —

the "someone" — for its own class interest and profit; and the second model, where money represents a receipt for goods and services — the "something" — produced by a state's people — the "someone" — who through the agency of their state issue that money to themselves, debt-free.

The problem of money is thus not even the *what* nor the *how much*, but the *who*, i.e., the who behind its issuance. The first model is a kind of "false alchemy" or a technology of black magic, for it is ultimately a technology to gain the mastery over the will, genius, and productive activity of a people, and ultimately, over the physical medium itself. And it is all based, as we shall now see, on a breathtaking series of historical, conceptual, and physical deceptions, and these can only exist so long as there are those unwilling to unmask the deception, or so long as there as those who acquiesce to that system and are unwilling to free themselves from its chains.

I.

Historical and Conceptual Background

"We shall have world government whether or not you like it —
by conquest or consent."
—James Warburg, son of Paul Warburg, February 17, 1950,
Testimony to the United States Senate Foreign Relations Committee

"Single acts of tyranny may be ascribed to the accidental opinion of a day, but a series of oppressions, begun at a distinguished period, and pursued unalterably through every change of ministers, too plainly prove a deliberate systematical plan of reducing us to slavery."
—Thomas Jefferson

"But when a long train of abuses and usurpations, pursuing invariably the same Object evinces a design to reduce them under absolute Despotism, it is their right, it is their duty, to throw off such Government, and to provide new Guards for their future security."
—Declaration of Independence of the United States of America

One

The Conspirators' Postwar Detente

⁘

> *"Since it is quite impossible to understand the history of the twentieth century without some understanding of the role played by money in domestic affairs and in foreign affairs, as well as the role played by bankers in economic life and in political life, we must take a glance at each of these four subjects."*
> —Dr. Carroll Quigley[1]

If, as the epigraph cited above suggests, it is "quite impossible" to come to an accurate assessment of the domestic and foreign policies of any state in modern times without an understanding of the role of money, then one would be equally justified in saying that it is quite impossible to understand ancient history — or *any* period of history — without understanding the role of money and its manipulation in social organization, policy, or science.

However, money, at least in modern times, is the result of an "alchemical" operation and a kind of "financial technology," the operation of transmuting nothing into something, in this case, of turning a mere entry on a bank ledger — a "nothing" — into a unit of commercial exchange — a "something." Likewise, alchemy is a "science" of transmuting base metals into gold, and that implies an underlying physics and technology to accomplish the act. To call *both* an alchemical operation is to imply the fact that beneath the magical operations of banking there lies a deep physics, and perhaps a profoundly misunderstood physics. In any case, these are the *conceptual poles* between which our story moves.

---

1 Carroll Quigley, *Tragedy and Hope: A History of the World In Our Time*, p. 54.

BABYLON'S BANKSTERS | 45

But there are two *historical poles* between which our story also moves.

A. A Hypothetical Scenario

Imagine for a moment that there is a Very High and extremely ancient Civilization. It has reached the apex of its social and scientific achievement. Then, in a paroxysm of madness and greed, it tears itself apart in a Great War. As the untold destruction reaches its climax and the great sophistication of its science and technology — the very science and technology it has used to wage its war — can no longer be sustained due to the enormous damage to its infrastructure, both sides see that all is lost, and they each conspire to salvage as much as they can of their science and technology by contriving an intricate symbolic language that can be decoded when civilization reaches a similar stage of scientific and social development. The strategy is one of long-term survival and eventual recovery, and each side fully realizes that the actual meaning of the symbols will most likely be lost in the short term. Nonetheless, they each conspire to create secret fraternities to maintain the symbols, pass them down, and to the extent possible, begin the work of their decoding and of the restoration of the very science and technology that led to their demise.

It is natural and reasonable to assume that, parallel to this activity, each side will take inventory of its remaining scientific and technological assets, and to secret them away for their potential rediscovery and reuse. Similarly, it is reasonable to assume that the "victors" of said war will take inventory of the "loser's" assets and technologies, and confiscate some, destroy what cannot be moved or otherwise used, and forbid to the vanquished any further development or deployment of such technologies.

If all this sounds vaguely familiar, like the activities of the victorious Allies or of vanquished Germans after World Wars I and II, then the reader will be correct.

But we assume a much more destructive war in our imaginary scenario. We assume the existence of a scientific and technological sophistication that would make all our modern instrumentalities of destruction seem but mere popguns. Similarly, we assume an extent of that mythical civilization that is truly cosmic in character, and a war of truly cosmic and interplanetary scope, of which all civilizations and wars that follow are but mere shadows on the long climb back to a similar peak of development. Similarly, the strategy of inventory-taking, of the secreting away of now lost technologies and sciences, and of the making of a complex of symbols and the founding of secret fraternities to preserve them, is far grander in scope than the activities of victorious Allies or vanquished Germans in recent memory. But note, it is the *scope* that is grander. The activities and strategies themselves, however, are not all that different.

Thus, at one pole, lost in the mists of prehistory and countless myths and ancient epics, an extraordinarily sophisticated and Very High Civilization blew itself apart eons ago in an interplanetary war, a war that cost them the very science that accounted for their fabulous wealth and power. In the wake of that cataclysm, what was left of that civilization purposed to maintain as much of that lost knowledge as possible, and, eventually, to recover all of it. Thus arose the secret societies, mystery schools, and civilizations that bore the stamp of its legacy, among which were the two that shall be our focus here: the Mesopotamian civilizations of Sumer, Babylonia, and Assyria on the one hand, and Egypt on the other. Certain elements within those societies quickly began the consolidation and extension of their power, and, as a result, were able to preserve at least a portion of the scientific legacy of their more sophisticated forebears, for that scientific legacy — distorted and blasted apart and fragmented as it was by that war — was indeed the very basis of their power.[2] The combatants of that war, both the "good guys" and the "bad guys," victors and vanquished, each to a certain extent went underground, the one seeking to preserve and recover the lost scientific bounty for the common good, and the other to recover it to make yet another bid for hegemony and world mastery. A sort of guerrilla and covert war between them — protracted for millennia — ensued, and, at times, even a détente was declared, for in the aftermath of that war, each to a certain extent needed the good offices and graces of the other to survive. Détente, and yet, covert warfare: this is the dynamic with which we must perpetually contend throughout this book.

At the other pole, in distinctly more modern times, but similarly, after yet another and more familiar war, from May 29th to May 31st, 1954, the gilded banking and corporate elites of Europe and North America met at a small hotel in Oosterbeek, Holland called the Bilderberg,[3] and ever since, this secretive annual gathering of the super-rich class of international banksters has been known by the name where their first meeting was held: The Bilderberg Group. And they — or rather, the international class of the very wealthy that they represent — show all the classic signs of having understood the historical and scientific lessons of their Egyptian and Babylonian predecessors, and more importantly, of *their* forebears of whom Egypt and Babylonia were but pale legacies.

And the timing of the meeting — 1954 — is in itself worth some commentary, for some allege that during 1954, President Dwight D. Eisenhower secretly met with an "extraterrestrial delegation" at a location in California. Others, however, maintain that Eisenhower was not really meeting

---
2 For the story of this very ancient cosmic war, see my *The Cosmic War: Interplanetary Warfare, Modern Physics, and Ancient Texts* (Adventures Unlimited, 2006).
3 Estulin, op. cit., pp. 19–20.

with extraterrestrials at all, but with very terrestrial Nazis at their secret Argentine headquarters.[4] Moreover, that time period — ca. 1954 — is the same period that the United States Air Force showed a distinct, but highly secret, interest in the free-energy research of one Nazi scientist in Argentina, Dr. Ronald Richter and what it portended for advanced propulsion and energy possibilities.[5] And, at the very same time period, discussion of anti-gravity physics disappeared in the open literature.[6] If one takes the unusually synchronous timing of these disparate events as something less than coincidental, and as somehow related to each other, then clearly one implication is that there is a deep interest within financial circles concerning deep physics. A struggle for financial, and technological, hegemony is subtly implied.

It is between these two unlikely *historical* poles — an extremely ancient one, and a very contemporary one — that this story also moves.

Very obviously, one is dealing here with something that, in modern times at least, looks very much like a conspiracy. And like all conspiracies, it has its "factions," groups vying for ultimate power that, for whatever reason and combination of circumstances, momentarily and temporarily unites them to make common cause to achieve the agreed upon goal of global domination and hegemony. The conspirators, whoever they are, thus appear at both poles of the story, agreeing to lay down their arms after a terribly destructive war, and to make common cause for the immediate future. They agreed, in short, to a coexistence, to a détente.

However, like all such détentes, this one is destined to break down. The disagreements, such as they are, may occur over the means and methods to achieve it, but ultimately and more importantly, they will occur — if they have not already — over which faction will ultimately lead it. In this, the modern circumstance resembles its "paleoancient" ancestor, the Very High Civilization that blew itself apart and then quickly established the legacy civilizations, the mystery schools and priestly fraternities and the monied class of merchants and bullion brokers that would seek to use one aspect of that lost science — the financial and economic "alchemy" — to recover the other aspect of it, the lost physics of the alchemical manipulation of the physical medium itself.

Thus, there are two other poles between which this story moves: the first being physics, and the second being the economics of high finance and its allied science of "social engineering." We have, then, four poles: two being temporal ones, widely separated in time, and two conceptual or disciplinary

---

4 See my *The Nazi International* (Adventures Unlimited Press, 2009), pp. 300–301.
5 Ibid., pp. 249–350.
6 Various authors and researchers have noticed this point, the most prominent among them being Nick Cook in his *The Hunt for Zero Point,* and Michael Schratt, in an excellent DVD presentation entitled *That's Classified: USAF Secrets Revealed* (Michael Schratt, Crystal Lake, Illinois).

ones, apparently widely separated by subject matter and method. For all the previous reasons, our story is partly an historical one, and partly a story about that strange intelligible world of very abstruse physics, of equally abstruse finance, and of the even more abstruse relationship between the two. It is thus above all an interdisciplinary story.

So, before proceeding any further, let us be absolutely clear and certain of what has been said, and what this book attempts to illustrate:

1) There is both in ancient and in modern times an "international monied class" deriving its power from its understanding and manipulation of the science of high finance, and from its monopolization of the money-creating power in its own private hands;
2) This class fought, at both ends of the historical pole, tremendously destructive wars. In the paleoancient[7] instance, this war was interplanetary in nature, and resulted in the fragmentation of the physics and economics components of a highly unified and integrated scientific worldview, in which those components were once united;
3) The result of that ancient cosmic war was the creation of legacy civilizations and secret organizations, in which that monied class swiftly established itself, on the basis of the retention within those societies' orbits of knowledge of at least some aspects of their ancestors' knowledge of the science of high finance and of its relationship to physics;
4) Thus established, this class has bent every effort toward the recovery of the physics component of their ancestral scientific legacy, and has blocked all attempts of those who throughout the millennia have sought to recover it and to *proliferate it among the masses*. The reason they have blocked all such attempts is quite simple, for once in the hands of the masses, their hegemony as a class would be irretrievably, and once and forever, broken.

So, in a sense, the paleoancient "Cosmic War" of which I have written extensively elsewhere, went underground. It became a covert war, a guerrilla war, and has been over the protracted millennia ever since a struggle between those who wish to maintain their power by reconstructing that lost unity of high financial economics and physics and monopolizing it for themselves to

---

7   I coined the term "paleoancient" in my first book, *The Giza Death Star*, to denote the extreme antiquity of this Very High Civilization, predating the known classical civilizations of Egypt, Sumer, the Vedic civilization, and China.

enslave mankind, and those who seek not only to reconstruct that lost unity but to share it, to "democratize" it among the great masses of humanity.

This book thus not only moves between two poles widely separated by time, or between two disciplinary poles — physics and high finance — apparently widely separated by subject matter and method, but it is also about the dynamics of that conflict, one which the banksters, for reasons that will become apparent in this book, are ultimately destined to lose, no matter what their extraordinary power or the Byzantine subtlety of their plans and cabals might be.

But surely, the reader will ask, to implicate the first meeting of the Bilderberg Group in this grand millennia-old story is going a bit too far?

On the contrary, the evidence points almost ineluctably toward that conclusion, but in order to see how, one must examine the Bilderberg meetings, and particularly the very first ones, in more detail. For the purposes of this examination, no one has examined and exposed those meetings in more detail than has Daniel Estulin, and he will accordingly be reviewed here.

B. Estulin's Study and Standard Interpretations

Estulin is a Russian émigré to Canada who first became interested in the Bilderberg group when he and others exposed the 1996 Bilderberg plot to dismember Canada, and have portions of that country "absorbed" by the United States.[8] Put on to the plot by a KGB agent whom Estulin calls "Vladimir," he and his erstwhile informant barely escaped with their lives when, prior to stepping onto an elevator, his informant held him back and pointed out that the elevator floor was missing![9] These two events compelled Estulin to uncover as much as he could about the secretive Bilderberg Group, and as a result, his study, *The True Story of the Bilderberg Group*, will be followed closely here, for the case I believe it makes is ultimately something very different than that Estulin, or other researchers of the Group, believes it to be.

In order to see how, though, it is necessary to see what Estulin actually uncovered about the Group.

1. Estulin's Sources

Estulin's book is amply documented with numerous photos, actual pictures of Bilderberg invitations, and so on. But how did he acquire such inside knowledge? His answer is quite simple: "I could not have done this

---

8   Estulin, op. cit., pp. ii, 6–7, 16–17.
9   Ibid., pp. 10–11.

without the help of 'conscientious objectors' from inside, as well as outside, the Group's membership."[10] A little further on he elaborates on who some of these sources were: the very hotel staff — the cooks, bellboys, waiters, cleaners and other staff the Bilderbergers took such pains to vet![11] And as has been seen, Estulin also received the considerable help of intelligence agents from Russia. This fact will play a major role in our conclusions toward the end of the book, in that it indicates that foreign intelligence services, particularly of those nations and power blocs not under Bilderberg influence, watch and analyze the plans of this self-appointed global elite very carefully and closely.

2. Protocols
a. Attendees

So who are the typical attendees of the Bilderberg meetings?
According to Estulin, they are what one might expect of such a gathering of a self-appointed world elite:

> (They) are annually attended by Presidents of the International Monetary Fund, the World Bank, and Federal Reserve; by chairmen of 100 of the most powerful corporations in the world such as Daimler-Chrysler, Coca-Cola, British Petroleum (BP), Chase Manhattan Bank, American Express, Goldman Sachs, and Microsoft; by Vice Presidents of the United States, Directors of the CIA and the FBI, Secretaries General of NATO, American Senators and members of Congress, European Prime Ministers, and leaders of opposition parties; and by top editors and CEOs of the leading newspapers in the world.[12]

Such a group, needless to say, reads like a who's who of the western world's power brokers:

> Every U.S. president since "Ike" Eisenhower has belonged to the Bilderberg Group, not that they have all attended the meetings personally, but all have sent their representatives. Another member is now ex-Prime Minister Tony Blair, as well as most of the principal members of the British government. Even Canada's high-profile past Prime Minister, Pierre Trudeau, was a member. Past Bilderberg invitees are Alan Greenspan, former chairman of the Federal Reserve; Hillary and Bill Clinton; John Kerry; Melinda and Bill Gates; and Richard Perle.

---

10   Ibid., p. ziv.
11   Estulin, op. cit., p. 10.
12   Ibid., p. xiv.

Also members are the people who control what you watch and read — media barons like David Rockefeller, Conrad Black (the now disgraced ex-owner of over 440 media publications around the world from *The Jerusalem Post* to Canada's newest daily, *The National Post*), Edgar Bronfman, Rupert Murdoch, and Sumner Redstone, CEO of Viacom, an international media conglomerate that touches virtually every major segment of the media industry. They have protected the secrecy of this secret society, and this may be why the name "Bilderberg" is new to you.[13]

With such a concentration of political, financial, and media power, it is understandable that the meetings have, for the most part, been kept fairly secret.

But this list does not really tell the full story, for considerable thought is given to the invitees to each year's gathering. For one thing, notes Estulin:

It is important to distinguish between active members who assist annually and others who are only invited occasionally. About eighty members are regulars who have attended for many years. Fringe people, who are invited to report on subjects related to their sphere of influence of professional and academic knowledge, are clueless about the formal structure behind the Bilderberg Group, and remain in absolute ignorance of the Group's greater goals and universal objectives. A select few are invited because the Bilderbergers think they may be useful tools in their globalist plan and are later helped to reach very powerful selected positions. One-time invitees who fail to impress, however, are cast aside.[14]

Note what we have here:

1) *Regular attendees* — people such as David Rockefeller and Henry Kissinger — representing the major financial and political interests of the group;

---

13 Estulin, op. cit., pp. 22–23. Estulin lists U.S. Secretary of Defense William Perry, Canadian Jean Chretien, the ubiquitous and sinister Henry Kissinger, David Rockefeller, George Soros, and various Belgian, Spanish, and Dutch royalty as attendees at the 1996 meeting that attempted to break up Canada. Q.v. p. 5. Estulin gives a further list of names on pp. 28–29. The breakup of Canada was to be accomplished via the mechanism of a "unilateral declaration of independence by Quebec" which would then, following the fragmentation of Canada, be absorbed piecemeal by the U.S.A. by the year 2000 (pp. 4–5). The contempt of the Bilderbergers for the perceptions of the general populace — perceptions they help to create and manipulate by their media organs — cannot better be illustrated than by this goofy scheme, for one cannot imagine French-speaking Quebecois faring any better in a union with the U.S.A. than they have with the Dominion of Canada, and probably would fare much worse.

14 Estulin, op. cit., p. 33.

2) *Occasional attendees* or their representatives, representing a similar makeup, but only invited as the situation demands; and
3) *Professionals or academics*, whose expertise is sought in certain situations, presumably to analyze and report on certain trends within the purview of their expertise, and presumably to make forecasts and "recommendations."

As we proceed, we shall see how each of these three groups break down in detail. For example, within each of these groups, a further breakdown occurs:

> Each country sends a delegation of, typically, three persons: an industry or business leader, a top-level minister or a senator, and an intellectual or chief editor of the leading periodical. The United States has the most participants because of its size. Smaller countries like Greece and Denmark are afforded, at most, two seats. The conferences usually consist of a maximum of 130 delegates. Two-thirds of the attendees are from Europe, and the rest come from the United States and Canada. (Mexican globalists belong to a less powerful sister organization, the Trilateral Commission.)
> One-third of the delegates are from government and politics, and the remaining two-thirds from industry, finance, education, labor and communications. Most delegates are fluent in English, with French as their second language of choice.[15]

Thus, a little more analysis is necessary. Note the following preponderances:

1) The Bilderberg Group is heavily weighted to a *European* point of view, though the United States has the largest single representation of any country sending delegates;
2) The Group is similarly weighted to the *private* sector rather than politics or government, with finance, industry, labor and academia being the most prominent members.

From this one may reasonably deduce that the Bilderberg Group, while ostensibly attempting to create a "united front" of these various internal groups and interests for a common cause, nonetheless contains within it two major factions, each of which has in turn two further factions within it:

1) The "European" Faction
2) The "North American" Faction (excluding Mexico).

---
15 Ibid., pp. 26–27.

Within these two main factions, one discerns two further factions:

a) The Political or Government Faction, stressing the role of public institutions of power and the bureaucracies that inevitably accompany them;
b) The Private Finance[16] Faction, stressing the role of the private monopoly of money creation and its dominant role influencing other major private sectors: labor, media, the academy, and so on.

These four internal factions will become quite important when we turn to consider the interpretation of the motivation for the actual *founding* of the Bilderberg Group later in this chapter, for as we shall see, this author departs significantly from the standard line of what its actual purposes, and power, may be.

Note one final thing from this analysis: *the private European faction is the dominant faction in terms of representation.* This is further evident from the fact that the Group, "from its inception, has been administered by a small nucleus of persons, appointed since 1954 by a committee of 'wise men,' which is made up of a European Chairman, both a European and a U.S. Secretary General, and Treasurer."[17]

That stated, however, there is one more significant detail that Estulin adds to this list of "typical Bilderberg attendees. Attendees, according to an unofficial Bilderberg press release, are drawn from a list of 'important and respected people who, through their special knowledge, personal contacts and influence in national and international circles, can amplify the objectives and resources of the Bilderberg Group.'"[18] There are, principally within the categories of "private" attendees, six further typical profiles:

1) industrialists;
2) financiers;

as we have already seen, but then the list gets interesting:

3) *ideologues*;
4) military professionals;
5) "professional specialists" such as lawyers, journalists, doctors, and so on; and
6) organized labor.[19]

---

16 By "finance" I mean henceforth the institutions, particularly the central private banks such as the Federal Reserve, the Bank of England, etc., that have in private hands the power to create the economic medium of exchange, debt instruments, credit instruments, and so on.
17 Estulin, op. cit., p. 27.
18 Ibid.
19 Ibid., p. 86.

What *sort* of ideologues, one might ask?

Many people have attempted to answer that question, and in doing so have attempted to analyze the true founding purposes and goals of the Group. Literature on the Bilderberg Group is therefore replete with documentation and speculation of the types of ideologues suspected to attend their meetings, encompassing a broad spectrum from Fabian socialists, globalists, exponents of World Government, animal and environmental activists, feminists, population control advocates, population *reduction* advocates, and so on.

There is one type of ideologue, however, that is *seldom* mentioned, and when it *is* mentioned, even the most perspicacious conspiracy researcher more often than not misses the significance of what it might indicate about the true founding purposes of the group, and the reasons for its very peculiar "union" of the four previously mentioned factions. In doing so, even these few researchers fall prey to the typical "standard interpretation" — which we shall examine a little later in this chapter — and attribute to the Group a monolithic structure and an almost omniscient and infallible power to accomplish its goals. In doing so, they tend to miss the recent and significant signals that there is serious factional infighting occurring within it and similar groups.

But how does this shadowy group manage to keep its existence and meetings secret? After all, a yearly meeting of the West's elite and powerful that has been ongoing since 1954 would surely cause *some* notice to break out, even in their own closely controlled media organs.

b. Vetting, Securing, and Announcing the Location

The answer, of course, lies in the extraordinary security that surrounds not only the meetings themselves, but more importantly, that surrounds the preparations for them, from the selection of the location, to the timing of the announcements of where the meeting is to be held.

> According to a source within the Steering Committee, "the invited guests must come alone; no wives, girlfriends, husbands or boyfriends. Personal assistants [translation: heavily armed bodyguards, usually ex-CIA and Secret Intelligence Service (SIS a.k.a. MI6)] cannot attend the conference and must eat in a separate hall. Not even David Rockefeller's personal assistant can join him for lunch. The guests are explicitly forbidden from giving interviews to journalists."
>
> To maintain their aura of hermeticism, Bilderbergers book a hotel for the duration of the conference, usually ranging from three to four days, with the whole building being cleared of all other guests by the CIA and local secret service to ensure complete privacy and

safety for the delegates. All drawings of the layout of the hotel are classified, staff is thoroughly vetted, their loyalty questioned, their backgrounds verified, and political affiliations checked. Any suspect ones are removed for the duration.

...

The host national government takes care of all the security concerns of the attending guests and their entourage. It also pays the costs of the military protection, the secret service, national and local police presence, as well as all additional private security to protect the intimacy and the privacy of the all-powerful world elite. The attendees are not required to follow the established rules and regulations of the host country, such as having to go through customs, carrying proper identification, such as passports, which are not required on Bilderberg visits. When they meet, nobody who is not on the "in" is allowed to come near the hotel. The elite often bring their own chefs, cooks, waiters, secretaries, telephone operators, busboys, cleaning staff and security personnel.[20]

As a further precaution, even though the timing of meetings is announced to invitees four months in advance, the actual location of the meeting is not made known to them until only one week prior to its occurrence. Finally, all documents distributed to attendees is marked personally for them, and stamped "personal and strictly confidential, and not for publication."[21]

Observe carefully what this implies, for such a gathering of the super-rich and powerful, with all the banking, media, and governmental contacts that this inevitably brings with it, is not only a Group that is gathering intelligence, circulating and exchanging information, but that it has its own internal classification system, for one might assume that the *contents* of such publications vary from delegation to delegation, if not between the individual attendees themselves. One may also assume that certain individuals receive rather more information — and thus have a higher "classification clearance" — than others.

c. *The Sessions and Their Rules*

Each meeting is divided into four daily sessions, "two in the morning and two in the afternoon," with the exception of Saturday afternoons which are left free for leisure activities.[22] When in session, the seating is by "rotatable

---

20   Estulin, op. cit., p. 25.
21   Ibid., pp. 30–31.
22   Ibid., p. 26.

alphabetical order," with "A's" sitting in front one year, and "Z's" the next, and so on.[23]

All session meetings follow the Rules of the Royal Institute of International Affairs, under which the contents of discussions may be shared, but the identities of individuals making statements, nor indeed, the identities of any individuals in attendance, may be revealed. Additionally, under these Rules, one may not reveal that one received this information at the Bilderberg Group's meetings.[24]

3. Goals and History: Bilderberg Shenanigans

As Estulin notes, the Bilderberg Group itself released a document in 1989 that stated their first meeting, held in 1954,

> "grew out of the concern expressed by many leading citizens on both sides of the Atlantic that Western Europe and North America were not working together as closely as they should on matters of critical importance. It was felt that regular, off-the-record discussions would help create a better understanding of the complex forces and major trends affecting Western nations in the difficult post-war period."[25]

But there was another goal from the outset that lay hidden within this desire simply to have "off-the-record discussions":

> Lord Rothschild and Laurence Rockefeller, key members of two of the most powerful families in the world, personally handpicked 100 of the world's elite for the secret purpose of regionalizing Europe, according to Giovanni Angelli, the now-deceased head of Fiat, who also said "European integration is our goal, and where the politicians failed, we industrialists hope to succeed."[26]

Bear the goal of European integration in mind, for it will become an important data point in the interpretation of the real founding motivations behind the Bilderberg Group in a moment.

Within a mere 14 years of that first meeting in 1954, however, Bilderberg goals — with the European Common Market well on its way to becoming the European Economic Union and finally the European Union — had taken on a somewhat grander vision:

---

23 Estulin, op. cit., 26.
24 Ibid., p. 27.
25 Cited in Estulin, op. cit., p. 23.
26 Ibid., pp. 23–24.

(A) comment made to the Bilderberger elite by George Ball during a presentation titled "Internationalization of Business" at the April 26–28, 1968, Bilderberg meeting in Mont Tremblant, Canada, provides a far more truthful and insightful glimpse into the Group's economic orientation. Ball, who was the Undersecretary of State for Economic Affairs under JFK and Lyndon Johnson, a Steering Committee member of the Bilderberg Group as well as a Senior Managing Director for Lehman Brothers and Kuhn Loeb Inc., defined what the new Bilderberger policy of globalization was going to be, and how it would shape the Group's policy.

"In essence," writes Pierre Beaudry in *Synarchy Movement of Empire,* "Ball presented an outline of the advantages of a new-colonial world economic order based on the concept of a *world company,* and described some of the obstacles that needed to be eliminated for its success. According to Ball, the first and most important thing that had to be eliminated was *the archaic political structure of the nation state.*"[27]

Estulin summarizes what this "world company" means in practical terms about Bilderberger goals. According to him, they want:

1) One International Identity;
2) Centralized Control of the People;
3) *A Zero-Growth Society;*
4) *A State of Perpetual Imbalance*;
5) Centralized Control of All Education;
6) Centralized Control of All Foreign and Domestic Policies;
7) Empowerment of the United Nations;
8) Western Trading Bloc (a NAFTA-like union of North and South America);
9) Expansion of NATO;
10) One Legal System;
11) One Socialist Welfare State.[28]

Bear in mind points 3 and 4 (Zero Growth and Perpetual Imbalance), as they will bear directly on our interpretation of the founding purposes for the Group a little later on.

These "modest" goals therefore influence the vetting process for who gets invited to Bilderberg meetings. Estulin states that the Group's Steering

---

27 Estulin, op. cit., p. 93, emphasis in the original.
28 Ibid., pp. 41–43.

Committee simply looks "for a One World Order enthusiast and (a) ... Socialist."²⁹ This is yet another important clue, one that will assume great significance as this book unfolds, for it means essentially that the Bilderberg Group's worldview is *an intellectually closed system*; dissenting or opposing points of view are not even represented in the Group's typical composition at any given meeting. As noted before, this gives a distinctively ideological characteristic to the Group's composition. After all, all of them are ideologues, and the ideology is, quite simply put, global domination by a self-appointed elite. The goal is simply "to rule the world with or without its consent, with guns or butter, without."³⁰

David Rockefeller, of course, is one of the principal movers and shakers in this cooperative elitist effort to enslave mankind, and in 1972, he and protégé Zbigniew Brzezinski founded the Trilateral Commission, a Bilderberg-like group that reached out to include Japan and the Orient within its ranks. The goal was to create an "international alliance that would create strategies and policies to consolidate the four pillars of power — 'political, monetary, intellectual and ecclesiastical' — under a central world government."³¹ The "ecclesiastical" component is a long and murky part of this story, and in fact too complicated to go into here as it would require a book of its own. Suffice it to say that the Rockefeller Foundation is the source of many grants to ardent ecumenical movement projects and has been a major supporter of the World Council of Churches. That being said, however, as will be seen in part two of the present book, the alliance between the banksters and a cynical "temple elite" has been a very long one, stretching back to ancient times.

In any case, at the 1991 Bilderberg meeting Rockefeller minced no words about his goals for a "supranational sovereignty of an intellectual elite and world bankers, which is surely preferable to the national auto determination practiced in past centuries."³² Given the inability of the Bilderberg Group to countenance viewpoints that do not fall into their "closed system" worldview, it is no surprise that it has taken active steps to remove from prominence those who espouse and/or act upon opposing views.

For example, the Bilderbergers opposed President Charles De Gaulle's creation of an independent French nuclear and thermonuclear arsenal because it threatened to make France too independent and nationalistic on the world geopolitical stage.³³ Estulin also maintains that the Group was behind the removal of Prime Minister Margaret Thatcher because of her opposition to

---

29   Estulin, op. cit., p. 24.
30   Ibid., pp. 8–9.
31   Ibid., p. 142.
32   Cited in Estulin, op. cit., p. 61.
33   Ibid., p. 20.

a currency amalgamation between Great Britain and the rest of Europe and the loss of national sovereignty this would inevitably mean for her country.[34] Estulin even presents evidence that Italian Prime Minister Aldo Moro was murdered by the Group for similar reasons, and that Henry Kissinger was implicated in the affair.[35] Finally, according to Estulin the real reason behind the whole sordid Watergate affair and the resignation of President Richard M. Nixon was because he opposed the General Agreement on Tariffs and Trade (GATT), a scheme that would lead inexorably to the NAFTA agreements and the eventual amalgamation of the U.S.A., Canada, and Mexico into a regional super-state.[36]

4. Tentative Conclusions and Implications: The Standard Interpretive Temptation

All these considerations lead many people to what I call the "standard interpretive temptation" of the Bilderbergers and similar groups,[37] namely, that they constitute a more or less monolithic group whose power and competence is scarcely resistible, and whose planning and subtlety is so comprehensive that they seldom fail in their goals, and most importantly, that they show little if any signs of serious fracture and fragmentation — of struggles for power — *within* their own circle. They are, in a word, a sort of *hive* or *herd*, and, adopting the hive and herd mentality for themselves, they seek to extend it to the rest of the world as a means of their own domination and hegemony.

This hive/herd mentality of the Bilderberg Group reflects itself in the way that it and all other such groups view the world economically, conceptually, physically, and politically, for in each of these four cases, they believe it to be a *closed* system. Therefore, it is necessary to have a basic working idea of what is meant by "closed systems" in each of these four senses.

By a closed economic system is meant the idea that the world's energy supply, upon which the financial system of the banksters is based, is founded upon non-renewable energy sources such as oil, natural gas, and so on. Hence, the system is closed since the whole premise of the system is *scarcity and non-renewability*. Because of this, "One Bilderberg objective is to de-industrialize the world by suppressing all scientific development, starting with the United States. Especially targeted are the fusion experiments as a future source of

---

34 Ibid., p. 50.
35 Ibid., pp. 50–51.
36 Ibid., 59.
37 Similar groups, i.e., the Royal Institute of International Affairs, or "Chatham House," The Milner Group, the Council on Foreign Relations, the Trilateral Commission, and so on.

nuclear energy for peaceful purposes."[38] Note the profound implications of this closed-system economic paradigm, for *it necessitates that the Bilderberg and similar Groups must resort to active measures to suppress the scientific development of new theories and their allied technologies that would shift the world's energy supply — and hence the financial system — to a new basis, and a basis no longer needing to be reliant upon their own monopoly financial power to create the medium of exchange and credit.* These active measures would include, but need not be limited to, the following:

1) Suppression of physics theories implying an "open systems" basis for energy, such as any hyper-dimensional theory, zero-point energy theory, and so on, that would lead to technologies tapping into them. Such theories are inherently *open systems theories* and the energy supplies they posit are virtually inexhaustible; hence the financial system would perforce undergo a dramatic metamorphosis, ending their financial money-and-credit creating monopoly, since finance itself would become an open system;
2) Promotion of physics theories in the public "marketplace of ideas" that are deliberately designed to "dead end," i.e., make practical technological access to such energy supplies "theoretically impossible" and thus close off scientific debate and discussion, i.e., physics itself is made to be a conceptually closed system;
3) Suppression of individuals, groups, corporations, or countries developing such energy systems, whether standard and conventional nuclear energy, to more exotic energy sources such as controlled fusion and so-called "free energy" devices;
4) Promotion of so-called "energy-efficient" technologies that are basically nothing but expensive "upgrades" of the current technologies and energy systems, leaving their monopoly financial power intact and unthreatened.

These considerations lead almost inexorably to a consideration of the next sense of "closed systems," to the "conceptually closed system."

As has already been indicated, the type of physics actively promoted by the Group will be deliberately designed to "dead end," i.e., it will be one of two things. It will either be a closed-system physics, or an open-system physics that *is impractical and even theoretically impossible to engineer for the foreseeable future.* Hence, the Group's own internal dynamics will not only be

---

38   Estulin, op. cit., p. 44.

conceptually closed to all such theories, but will extend itself across the range of human experience, and conceptualization, since a practically engineerable open-systems physics implies the democratization not only of energy, but its corollaries, finance and political power, across a very broad spectrum of people, as a greater mass of people would be lifted *up* to greater wealth, freedom, and prosperity. The Group, then, evidences the effect of its own closed-system physics, by advocating closed and monopolized systems of finance and politics. This was reflected in the fact that the Group recruits only those who share its view and goals, as we have seen.

And this highlights its first major weakness, for being committed rather unscientifically to such a closed system in which progress is to be shunned and suppressed, the Group shows that it is incapable of adaptability to changing circumstances that are not under its own control. In short, it cannot *evolve* and is therefore destined to stagnate, wither, and die.

This fact evidences a further problem that the Group has, and it is one to which I have averred in previous books and radio talk show interviews, and that is the "proliferation problem." Estulin outlines this problem in the following fashion:

> Why is nuclear energy hated so much by the New World Order? According to John Coleman, a former British MI6 secret agent, nuclear power stations generating abundant cheap electricity are "*The key to bringing Third World countries out of their backward state. With nuclear energy generating electricity in cheap and abundant supplies, Third World countries would gradually become independent of U.S. foreign aid, which keeps them in servitude, and begin to assert their sovereignty.*
>
> Less foreign aid means less control of a country's natural resources by the IMF, and greater freedom and independence for the people. It is this idea of developing nations taking charge of their own destinies that is repugnant to the Bilderbergers and their surrogates.
>
> This was confirmed on page 13 of the Bilderberg 1955 General Report: "In the field of atomic energy, scientific discovery was continually overtaking itself.... It could not be excluded that the scientists would put the bomb into the hands of more and more people and so 'the atomic bomb would become the arm of the poor.' The same applied to the development of atomic energy for peaceful purposes, where we had almost to foresee the unforeseeable."[39]

---

39 Estulin, op. cit., pp. 44–45.

In other words, part of the concern of the Bilderbergers in the decade or so following World War II, was the apparent ease by which atomic energy — and hence atom bombs — could proliferate beyond their ability to control. It would, to put it mildly, be a political nightmare, as smaller nations could conceivably hold the rest of the world under a kind of nuclear blackmail. In short, *some* of their reasons for scientific suppression were indeed altruistic.

However, there all altruism ends, for a virtual Western monopoly on such weapons at that time in the immediate postwar period also gave the Group a final "threat of last resort" against the very people and world it sought and still seeks to enslave. And this brings us back to the other problem: open-systems physics and energy sources. As we shall see subsequently in this book, such systems are relatively easy things to engineer, and moreover, such a physics, if weaponized, would potentially be capable of making a hydrogen bomb look like a firecracker. The Group must therefore seek not only to suppress public exploration and development of this physics from a proliferation standpoint, but it must also seek to develop it privately for itself, giving it yet another monopoly on a technology of hegemony. This too, as we shall see, is yet another link to the ancient past.

There is one final consequence of such a closed-system view of physics, economics, and finance, and that is that it leads to perpetual war,[40] since ever-dwindling resources are being sought by ever more competing interests. Nor is the Group immune from this, since scarcity will inevitably bring about severe factional infighting, as one faction demands more resources than the projected "World Government" or "World Company" wishes to provide it.[41] Eventually, within such a closed-system physico-economic order, the system must devour the world and finally itself.

It is, by its own lights and philosophy, inevitably *doomed.*

If anything emerges from these brief considerations it is this: *in the emerging global community, an open-system physics is ultimately the key to open systems of economics, to open conceptual systems, and to open systems of politics and finance.* This deep and profound relationship between physics, finance, and political order is, of course, not unknown to the Bilderberg Group or other such groups, as their own active suppression of such physics and cognate technologies attests.[42] The real question that shall preoccupy us throughout the remainder of this book is exactly why that relationship should even exist, to outline its salient features, and to examine certain aspects of how it has manifest itself in history.

---

40 Estulin, op. cit., p. 95.

41 Estulin mentions the case of the economic rape of Argentina by the banksters, for the simple reason that nuclear Argentina was supplying much of Mexico's own energy requirements, and doing so without going through the oil spigots controlled by the Group and its oilmen. Q.v. p. 45.

42 This active suppression and its connection to the banking interest will be explored more fully in subsequent chapters.

C. Unusual Guests at the First Meeting, and the Alternative Explanation: Detente

As previously indicated, the standard interpretation of the Group and its surrogates as a monolithic, omnicompetent and omnipotent bloc does not take into account some very significant facts that bear directly upon the interpretation of those early Bilderberger meetings, including especially the first ones ca. 1954–1960. These facts may be distinguished into three classes:

1) The presence of two unusual guests, with prominent Nazi backgrounds, at the first meetings;
2) The wider context of postwar survival of a Nazi International and its financial and political capital and objectives;
3) The occurrence of events — subsequent to the first Bilderberg meeting — indicative of some sort of understanding between the Anglo-American banking elite and the Nazi International, each of which had their own globalist agenda for world domination.

Each of these points must be examined in some detail in order to appreciate their significance for an understanding of at least one of the probable purposes for the establishment of the Bilderberg Group.

1. Two Prominent Guests with Nazi Backgrounds

Most researchers on the Bilderberg Group have noticed the presence of Prince Bernhard of the Netherlands as the instigator and chairman of one of its first meetings in the 1950s, and some, Jim Marrs for example, have also noted that Bernhard was a former SS officer, as well as a vice president for the notorious and gargantuan German chemicals cartel I.G. Farben.[43] But there is another significant figure present in those first meetings as well, and it is the combination of these two men that indicate there may have been yet another hidden purpose in the founding of the Bilderberg Group.

That man is Deutschebank chairman, Sovereign Military Order of Malta member, and I.G. Farben, Daimler-Benz, and Siemens board of directors member, Dr. Hermann Josef Abs.[44]

In addition to these directorships on Germany's most powerful and

---

43 See Jim Marrs, *The Rise of the Fourth Reich: The Secret Societies that Threaten to Take Over America*, p. 48.
44 Estulin, op. cit., pp. 200–201. Estulin actually reproduces the Bilderberger documents on these pages. For Abs' membership in the Sovereign Military Order of Malta (SMOM) and directorship on the board of I.G. Farben, Daimler, and Siemens, see http://moversandshakersofthesmom.blogspot.com/2008/08/hermann-abs.html, p. 2.

largest bank and some of its largest corporations including Siemens and I.G. Farben, Abs, during his tenure as a partner of the private bank of Delbruck, Schickler and Co., located in Berlin, became a close associate of Nazi Party *Reichsleiter* and financial genius Martin Bormann, since Abs' private bank was the very bank that managed the accounts for the Reichschancellery, and thus paid Adolf Hitler's salary as Chancellor![45]

2. The Wider Context of the Postwar Nazi International and The Name "Bilderberger" Group

But there is a deeper implication to Abs' presence at the Bilderberg meetings, along with fellow I.G. Farben officer Prince Bernhard, than even this "friendly" relationship with Martin Bormann would indicate. This deeper implication is clearly suggested by the role that Deutschebank played in moving the massive amount of funds overseas that were a component of Martin Bormann's "Strategic Evacuation Plan," first outlined for the Nazi and German corporate leadership at a top secret meeting held in Strasbourg, France, in August of 1944.[46]

As I wrote in one of my previous books, *Nazi International,* seven things emerged as a consequence of this meeting:

1) Bormann basically confiscated any foreign reserves in the possession of German corporations and placed them under the control of the Nazi Party;
2) These funds in turn would be disbursed to corporations in aid of their fulfillment of the goals of the conference;
3) One primary goal was to establish "research bureaus" whose purpose, following the I.G. Farben N.W. 7 model, was to gather intelligence and steal foreign research as well as to conduct their own research;
4) Since these technical offices were to have their own Nazi Party liaison officer reporting directly to Bormann, it is clear that Bormann *intended* for his Nazi International to conduct not only espionage but *research* in its own right, under departments — the "technical offices" — of German corporations only very loosely connected to them;
5) Each such research bureau was to be established under a cloak of other activity, such as "investigating water resources";

---

45 http://moversandshakersofthesmom.blogspot.com/2008/08/hermann-abs.html, p. 2.
46 For the story of this meeting and its implications, see my *Nazi International: The Nazis' Postwar Plan to Control Finance, Conflict, Physics, and Space* (Adventures Unlimited Press, 2008), pp. 63–83.

6) The capital flight program was designed to see to it that Allied authorities did *not* get the most valuable German scientists, technicians, or knowledge, as represented in patents, all of which were to be transferred overseas to safe havens, thus indicating that Bormann, in the light of all the previously enumerated considerations, intended for the Party to continue *an independent line of scientific research under its own control,* all the while managing and sharing any information gained through its espionage activities with the involved German corporations, as the Party and Bormann saw fit.

And all this implies something else, something very obvious, and very significant:

7) Martin Bormann *fully intended to survive the war in order to coordinate all these activities.* [47]

Note the significant point of establishing "research bureaus" under the direction of Nazi Party liaison officers who, in turn, gathered intelligence, oversaw ongoing postwar Nazi research activities, and who reported directly to Bormann! By extension then, the presence of SS Officer Prince Bernhard as not only a founder of the Bilderberger Group, but as the first chairman of its first meeting, and the presence of Bormann's banker, Dr. Hermann Josef Abs, in later meetings, himself a Farben director along with Prince Bernhard, is a strong indicator that *the Bilderberg Group may have been founded as yet another "liaison office" for Bormann's Nazi International, allowing it not only the opportunity to sit down behind closed doors and "exchange information" with its chief rival for power — the Anglo-American banking elite — but to maintain a constant flow of intelligence on its activities and goals.*[48] It was, in short, a marriage of convenience to move and launder massive amounts of money through the Anglo-American elite's corporations and banks. And as with all such deals with the devil, Bormann's money also carried with it the implied threat that someday the "markers would come due."

There is some corroborative evidence for this interpretation even in Estulin's book itself:

---

47 Joseph P. Farrell, *Nazi International: The Nazis' Postwar Plan to Control Finance, Conflict, Physics, and Space* (Adventures Unlimited Press, 2008), p. 76.

48 Abs' easy access to West German Chancellors Konrad Adenauer, Ludwig Erhard, and Kurt Georg Kiesinger (a former member of Dr. Josef Göbbels' Propaganda Ministry!) may account for the fact that Bormann and his Nazi International were such hot political topics for West German politicians and prosecutors. Indeed, those prosecutors often would complain of their feeling that some hidden hand was directing the West German government, and inhibiting their efforts to bring Bormann and other top Nazis to justice. Abs, given his close contact before, during, and after the war, may have indeed been the go-between between Bormann and the postwar German governments.

Most reports contend the original members named their alliance the Bilderberg Group after the hotel where they made their covenant. Author Gyeorgos C. Hatonn, however, discovered that German-born Prince Bernhard was an officer in the Reiter SS Corp in the early 1930s and was on the board of an I.G. Farben subsidiary, Farben Bilder. In his book, *Rape of the Constitution; Death of Freedom*, Hatonn claims Prince Bernhard drew on his Nazi history in corporate management to encourage the "super secret policy-making group to call themselves the Bilderbergers after Farben Bilder, in memory of the Farben executives' initiative to organize Heinrich Himmler's "Circle of Friends" — elite wealth-building leaders who amply rewarded Himmler for his protection under National Socialist programs, from the early days of Hitler's popularity through to Nazi Germany's defeat. The royal Dutch family discreetly buried this part of Prince Bernhard's background when, after the war, he became a top official in Royal Dutch Shell, a Dutch-British conglomerate. Today, this rich European oil company forms part of the inner circle of the Bilderberg elite.[49]

There, however, Estulin leaves matters, never to return to the potential implications this information signals.

These potential implications are signaled, once again, by Abs' presence as a Bilderberger, and to see what it is, we have to look a little closer at the postwar relationship between him, Bormann, and the North American faction of the Bilderberg Group that is implicated in a little-known incident.

3. An Event Indicative of an Understanding between the Anglo-American Corporate Globalist Elite and the Nazi International

This incident is a clear indicator that the Bilderberg Group's early founding was for the purposes of striking a kind of "détente" between the two old rivals, the Nazi International and the Anglo-American elite, since both shared similar goals, and both had access to a *lot* of money. The incident was disclosed by longtime CBS journalist and Ed Morrow associate Paul Manning, in his book *Martin Bormann: Nazi in Exile.* Manning recounts a little-known effort by U.S. Supreme Court Justice Jackson, who had obtained fame as America's chief prosecutor during the Nuremberg War Crimes Tribunals, and President Harry S. Truman, to track down the Nazi Party *Reichsleiter.* I recounted this effort in my book *The Nazi International* as follows:

---

49 Etsulin, op. cit., p. 20.

Paul Manning, a journalist and longtime associate and friend of the famous CBS newsman Ed Morrow, mentioned Bormann's presence in Bariloche province, but did not connect his appearance with the Richter fusion project. Manning begins by noting that President Truman himself became involved in the hunt for Bormann in 1948, three long years *after* Bormann supposedly died trying to escape Berlin, according to the "official standard" history:

On June 16, 1948, President Truman became involved in the hunt for Martin Bormann. Robert H. Jackson, who had once taken a leave from the Supreme Court to serve as U.S. chief prosecutor at the Nuremberg trials, wrote to the president that a quiet search should be made by the FBI for Bormann in South America.

"My suggestion, therefore," he wrote, "is that the FBI be authorized to pursue thoroughly discreet inquiries of a preliminary nature in South America.... I have submitted this summary to Mr. Hoover and am authorized to say that it meets with his approval. You may inform him of your wishes directly or through me, as you prefer."[50]

The choice of the FBI to do the investigation is perhaps significant. Under American federal law after Truman signed the National Security Act of 1947 that created the CIA and NSA, the FBI was restricted to intelligence operations on American soil. All operations on foreign soil fell under the jurisdiction of the CIA. So why would Jackson have urged President Truman to undertake an investigation in South America through the FBI? After all, Truman himself signed the law into existence and knew full well its contents.

The answer lies once again in the fact that the CIA, which, under Zurich OSS station chief Allen Dulles during the war, had negotiated a highly secret deal with the head of German military intelligence on the Eastern Front, General Reinhard Gehlen, to turn over — lock, stock, and Nazi — Gehlen's entire network to the nominal oversight of the American intelligence community, leaving Gehlen in charge of its actual day-to-day operations. In other words, before the ink was even dry on the National Security Act of 1947, the CIA's "civilian character" had already been compromised in the most egregious way, since its entire operational desk in Eastern Europe and the Soviet Union, and to a great extent elsewhere as well, was staffed by a network of Nazis! Needless to say, then, the CIA would not be the best instrumentality to investigate Bormann's possible presence in Latin America! Jackson's request also suggests something even darker:

---

50 Paul Manning, *Martin Bormann: Nazi in Exile* (Lyle Stuart, Inc. 1981), p. 204.

that both he and Truman *knew* of the extent of Nazi penetration of the CIA, and therefore of its untrustworthiness as an investigative agency for postwar Nazi activity.

In any case, as Manning observes:

"The presidential authorization was given, and John Edgar Hoover assigned the investigation to his most experienced and skillful agent in South America, who proved that he was just that by eventually obtaining copies of the Martin Bormann file that were being held under strict secrecy by Argentina's Minister of the Interior in the Central de Intelligencia. When the file (now in my possession) was received at FBI headquarters, it revealed that the Reichsleiter had indeed been tracked for years. One report covered his whereabouts from 1948 to 1961, in Argentina, Paraguay, Brazil and Chile. *The file revealed that he had been banking under his own name from his office in Germany in Deutsche Bank of Buenos Aires since 1941; that he held one joint account with the Argentinian dictator Juan Perón, and on August 4, 5, and 14, 1967, had written checks on demand accounts in First National City Bank (Overseas Division) of New York, The Chase Manhattan Bank, and Manufacturers Hanover Trust Co., all cleared through Deutsche Bank of Buenos Aires.*"[51]

Bormann's wartime contact with Deutsche Bank bigwig Dr. Hermann Abs, and the connections to the international financial interests of Morgan and Rockefeller via their big New York banks, was paying big dividends, for Bormann quite apparently did not even have to *hide* his identity, but could sign and cash checks under his own name through some of the largest and most well-known banks in America as late as 1967![52]

Observe carefully the dynamic here:

1) Bormann has accounts *after* the war in large American banks representing the Rockefeller-Morgan interest;
2) He is able to draft checks on those accounts and cash them under his own name 22 years after the end of the war;
3) Those checks are cleared through his bank, Deutschebank, through its local Buenos Aires office in Argentina!

We have thus a Bormann-Abs-Rockefeller network, and this clearly implies that the two financial elites, represented by their respective delegations to

---

51 Manning, *Martin Bormann: Nazi in Exile,* pp. 204–205, emphasis added.
52 Farrell, *The Nazi International* (Adventures Unlimited Press, 2008), pp. 302–304.

those early Bilderberg meetings, have reached some sort of *modus vivendi*. The question is, why?

4. Closed Systems, Open Systems, and Further Possible Reasons for the Détente

A clue to the answer is afforded in part by the fact that the immediate postwar goals of the Nazi International and the Anglo-American elite are one and the same: the creation of a European federation under German economic dominance. Jim Marrs, in his bestselling book *The Rise of the Fourth Reich*, puts it this way:

> After the war a devastated Europe looked to Germany for economic leadership. The economic steps taken that became the Common Market took the shape of prewar Nazi plans. "(S)omehow the Gernans had the answer originally in 1942 when they were melding the economic institutions of the Continent into their own design," noted Manning.
> It is interesting to note that the present European Union (EU) began as merely economic measures.
> …
> The European Economic Community, better known as the Common Market, was established in 1957 by the Treaty of Rome.… George McGhee, a member of the secretive Bilderberg Group and former U.S. ambassador to West Germany acknowledged that "the Treaty of Rome, which brought the Common Market into being, was nurtured at Bilderberg meetings."[53]

And needless to say, the ultimate goal of world domination was a goal shared both by the North American Faction and the Nazi International, and hence, both could afford to make common cause, to come to an agreement of "détente" and "coexistence" at those first Bilderberg Group meetings.

But there is yet *another* reason such a détente was not only sought by both factions, but in a sense, altogether necessary: money. The rebuilding of Europe, not to mention its integration, would require money, and lots of it, and there were two groups — The North American Bilderberg faction, and the Nazi International — that had it in abundance. For Bormann to "grow" the substantial monies that his Nazi Party had plundered from Europe, he would have to have access to the international banking community and its

---

[53] Jim Marrs, *The Rise of the Fourth Reich: The Secret Societies that Threaten to Take Over America* (William Morrow, 2008), pp. 215–216.

banks in Britain and North America. Conversely, such vast influx of liquid cash and hard commodities such as gold, diamonds, and platinum, which Bormann's Nazi International had in abundance,[54] would constitute a reserve that would greatly expand the Rockefeller-Morgan-Rothschild banks' ledger credit-making ability, enriching them in the process.

It was, inevitably, a "détente" made in banksters' heaven.

And Bilderberg was the result.

There is, however, a final reason for the détente, one that once again demonstrates the deep connection of finance and physics.

For Bormann's Nazi International to maintain this détente for such a long period after the war — after all, according to Manning and the Argentine intelligence service, he cashed checks on Rockefeller-Morgan banks via Deutschebank in 1967 — he would have to have had powerful coercive "leverage" with which to ensure that the Anglo-American faction lived up to its part of the deal. Part of this no doubt came from the implicit threat afforded by the vast intelligence resources available to him: General Gehlen's old German military intelligence unit, now the nucleus of West Germany's *Bundesnachrichtendienst,* the Nazi Party's own significant intelligence network, and so on. The mere existence of such a network would convey the threat that if the other faction did not behave according to their agreements, its leaders could be very easily assassinated.[55] There is, however, another bit of leverage Bormann's Nazi International controlled, and it was one piece of leverage that its rival North American faction dearly wished to possess: alternative physics.

Over the course of four previous books on wartime Nazi secret weapons research — *Reich of the Black Sun, The SS Brotherhood of the Bell, Secrets of the Unified Field,* and *The Nazi International* — I have detailed the story of the Nazi Bell device, a hyper-dimensional physics device being researched under the auspices of the SS departments *Entwicklungstelle-IV* (Development Area IV), *Forschung, Entwicklung, und Patente* (Research, Development, and Patents), and SS General Hans Kammler's super-secret weapons black projects think-tank, the *Kammlerstab.*[56] As the mission brief of the *Entwicklungstelle IV* was to develop free energy and to make Germany independent of foreign oil,[57] and as the latter two departments were responsible for the Bell itself (a device I believe to have been designed for a threefold purpose: to investigate that same "free energy," to explore the possibilities of advanced field propulsion concepts, and finally, as a weapon that had the potential to make even the

---

54 See my *Nazi International,* pp. 64–66, 82–83, 176–177.
55 Not to mention the blackmail potential offered by the "détente deal" itself!
56 See my *Reich of the Black Sun,* pp. 99–107; *SS Brotherhood of the Bell,* pp. 144–148, 170–171.
57 See my *SS Brotherhood of the Bell,* pp. 170–171.

largest thermonuclear bombs pale by comparison for destructive potential),[58] this device and the alternative physics it represented and embodied were a clear threat to the closed-systems approach *of at least the North American* faction within the Bilderberg Group and similar groups. As I also detail in *The Nazi International,* Bormann saw to it that while the Allies and Soviets received an almost equal and stalemating division of the technological spoils of the Third Reich, this device he retained for the Nazi International, and continued a line of independent investigation of its physics.[59]

This independent investigation was headquartered in Argentina, near the remote city of San Carlos de Bariloche in Rio Negro province, some 900 miles southwest of Buenos Aires. Ostensibly a project investigating techniques for controlling thermonuclear fusion, it had evidently achieved some success, as dictator Juan Perón made an announcement in 1951 to this effect.[60] Further investigation, however, revealed that the ultimate goal of the project was far beyond this, as the project's head, Dr. Ronald Richter, viewed controlled fusion and plasma processes as a way to manipulate and the zero point energy, the fabric of space-time itself.[61]

For our purposes here, however, it is important to note that the United States shows an interest in Richter's project some four years *after* it was shut down by Bormann and his Nazi International.[62] This fact, plus the fact, according to Bell researcher Igor Witkowski, that after the war the United States made every effort to reconstruct the personnel team that worked on the Bell,[63] and that it also even went so far as to launch a military commando raid into Czechoslovakia in 1946 to recover documents doubtless relating to General Kammler's secret weapons group,[64] is a strong indication that the Anglo-American faction *knew* in a general fashion of the existence of this alternative physics in postwar Nazi hands, and was attempting to recover as much of it for themselves as possible, and break Bormann's monopoly hold over it and restore a "balance of power," albeit a covert one, between the two factions.[65] Needless to say, there is also every indication that the Soviets

---

58 See my *SS Brotherhood of the Bell,* pp. 141–308, and *Secrets of the Unified Field,* pp. 238–288.
59 See my *The Nazi International,* pp. 85–136; 348–350, 382.
60 Ibid., pp. 249–274.
61 Ibid, pp. 275–350, particularly pp. 316–317; 343.
62 Ibid., pp. 278–298.
63 Witkowski, *The Truth About the Wunderwaffe,* p. 260,
64 See my *Secrets of the Unified Field,* pp. 296–312.
65 Yet another factor is worth considering in this respect, and that is my interpretation of the Roswell event on the basis of the internal evidence of the MJ-12 documents as having been perhaps the crash of something *Nazi.* (Q.v. my *Reich of the Black Sun,* pp. 274–330; and *The SS Brotherhood of the Bell,* pp. 311–384.) If this be true, then the North American Faction, through the agency of its surrogate, the U.S. government and military, had yet another clue indicating the *postwar* Nazi possession of this physics, and its continued development of it. As such, it had yet another impetus to seek and recover as much of the original Nazi documentation and personnel as it could, and, failing that, to embark on its own programs

and Allies began a crash program after the war to investigate similar lines of physics for themselves.[66]

This indicates a possible final reason for the détente: Bormann's Nazi International could literally dangle portions of that physics and its cognate technologies before his erstwhile allies' noses, and dole it out piecemeal, if he so chose, in return for "favors." And this hints at a potential internal dynamic within the factions represented by the Bilderberg Group, for each faction would seek not only to suppress *public* development of such physics and technologies, but also seek to inhibit *the other faction's* development of it as well.

5. An Analogy of the Détente: The 1939 Nazi-Soviet Pact

An apt analogy of this situation is afforded by the Nazi-Soviet Pact of 1939, the pact that secretly divided Poland into Nazi and Soviet "spheres of influence" and made World War II possible. Both regimes being what they were, both took cynical advantage of the geopolitical situation, fully knowing that at some future time the two nations would probably come to blows. The pact was simply a "détente," a "coexistence" wherein both sides agreed not to go to war with each other, allowing a period of time during which each could build up its forces to do precisely that!

For reasons that were in part outlined in the prologue and to be explored subsequently, this author believes that within the last decade or so, there are distinct signs that that the postwar détente within the Bilderberg Group is breaking down, and that there is quiet, though very serious, covert war being waged between the various factions of that international corporate elite.[67]

6. Closed Systems, Globalization, New Energy Sources, and Outer Space

But what are the practical implications of this "closed system" approach to physics and economics?

For the conspirators adopting them and imposing them on the rest of the world, they are very simple. In the first instance, that of "globalization,"

---

to acquire the physics and technology for itself.

66 For the Soviet interest, see my *SS Brotherhood of the Bell,* pp. 203–236, and *The Philosophers' Stone* (Feral House, 2009), section four; for the American pursuit, see Paul A. LaViolette's excellent study of off-the-books projects in antigravity field propulsion in the U.S.A., *Secrets of Antigravity Propulsion: Tesla, UFOs, and Classified Aerospace Technology* (Bear & Co., 2008), pp. 42–259.

67 As yet another sign that this may be the case, one should also consider the allegation of authors Richard C. Hoagland and Mike Bara in their *New York Times* bestseller, *Dark Mission: The Secret History of NASA,* wherein they present evidence that by the time of the Apollo lunar missions, the Nazi faction had become the dominant and most influential group within the Agency, competing against the other two factions: Magicians, and Masons, both groups with strong ties to the Anglo-American establishment.

a closed global economic order will eventually devour itself, and therewith, the conspirators themselves. Knowing this, they have two possible alternatives: (1) development of new energy sources which they can monopolize, a risky prospect as we shall see, or (2) they can "open" the system up a bit, and venture into outer space, creating a potential "competitor," while simultaneously promoting factional and regional imbalances on the earth which they can manipulate and control, and maintain the illusion that the economy is perking along nicely (as they have done in the past). Both components of this stratagem are themselves fraught with risks as well, for in promoting regional imbalances and factional infighting, they run the risk that they will eventually lose control of them, and that the unitary closed global system will fragment, and their power be correspondingly diminished.

D. Conclusions and Implications

So what may one conclude from this examination of the real purposes behind the Bilderberg Group?

As we have seen, the following points now evidence themselves:

1) There is a deep and profound connection between the closed-systems approach of finance and politics advocated by such international groups of the corporate elite, and closed-systems physics;
2) This connection necessitates that any internal factions within it seek to suppress the open and public development of such physics, as well as to monopolize it for themselves and deny it to other factions;
3) Any such group must also seek to extend its influence over the other areas of power: academia, labor, and, as we saw from David Rockefeller's remarks, religion. This will figure prominently in part two in our examination of why ancient temples functioned also as centers for the then banking class;
4) In the wake of World War II, an apparent détente was struck between two major factions — the Anglo-American, and the Nazi International or European faction — as a result of the pressing need for channeling vast amounts of Nazi plunder back into Europe, and as the North American Group and the European group each used the meetings to gather intelligence on the other, and keep abreast of its activities.

With these observations, we have seen at one end of the historical pole the interrelationships of closed and open systems, of the banking elite, and of the necessity for it to suppress the development of open-systems physics that would lead to open financial systems. We have moreover suggested that there is serious factional polarization and tension within that class hiding just beneath the surface, and able to break out into open conflict given certain circumstances.

It is now time to press our examination a little further back into history, toward the Great Depression, and a hidden, little-known and little appreciated legacy of one of America's most unpopular Presidents. Thereby we will be afforded the perspective by which to perceive quite clearly the interrelationships of physics and economics itself, and to track that relationship even further back into history.

Two

HOOVER'S HIDDEN LEGACY AND GIFT
INDICATORS OF A LOST SCIENCE

∴

All our chart can tell us is that, failing a revolution of some sort, the pattern as established will presumably prevail."
—Edward R. Dewey and Edwin F. Dakin[1]

No President of the United States of America has gone down in history more lamented, and even excoriated, than has Herbert Hoover, whose twin misfortunes were first to have presided over the country at the beginning of the Great Depression, and secondly, to have been such an ardent free-market ideologue that he genuinely thought there was nothing he nor the government could nor should do to alleviate the suffering of those who had lost their savings, farms, and homes during the crisis. In this he came the closest of any American chief executive to the attitude of Marie Antoinette who, upon hearing of the plight of France's poor being unable to purchase adequate food and bread, is reported to have said merely, "Let them eat cake." His presidency thus came to epitomize the popular perception that the U.S.A. was a government of Wall Street, by Wall Street, and for Wall Street. Marie Antoinette, of course, lost her head over the matter. Hoover managed to retain his head, but was expelled from the presidency in 1932 in an election that swept Franklin Delano Roosevelt into office. But Hoover did leave one enduring legacy of his attempt to find a genuinely long-term solution to the problem of "boom-and-bust" business cycles that had plagued

---

1 Dewey and Dakin, *Cycles: The Science of Prediction* (Kessinger Publishing: reprint of the Henry and Holt Company book of the same title, 1947), p. 28.

American history, and it is a legacy few people even know exists. And even among the few that do know of its existence, even fewer of them know its true significance.

That legacy was Edward R. Dewey, and *his* legacy, the Foundation for the Study of Cycles.

Dewey's interest in cycles began when Hoover, beset by the misery the Great Depression was inflicting on so many Americans, wanted to know the cause of it, and assigned the U.S. Department of Commerce the task of finding it out. The assignment fell to Dewey, who was at that time the Chief Economic Analyst of the Commerce Department.[2] Seeking out the assistance of various economists, he lost faith in standard methods of economics when they could not provide him with coherent and consistent answers and models. Having access to the vast amounts of information available to him from the Commerce Department, Dewey adopted a novel approach. Rather than concerning himself with the *whys and causes* of depressions, he decided to study *how* economic and business cycles occurred. He decided, in other words, to study the *what*, and not the *why*,[3] and in doing so, he discovered something very significant.

He discovered that cycles, *waves* of behavior, existed in almost all aspects of economic and human social life, from the prices of pig iron to human emotions themselves. According to the Foundation for the Study of Cycles' website itself, Dewey learned of a "1931 Canadian conference on biological cycles. Under the guidance of Dewey and the conference leader, Copley Amory, the conference's Permanent Committee was organized into the Foundation for the Study of Cycles, and *its scope was enlarged to encompass all disciplines.*"[4] Having access to this expanded database led to the Foundation's establishment in 1941, and it has continued to gather hard *empirical* evidence on all manner of cycles in all possible disciplines ever since. The work of the Foundation continued in Dewey's tradition, gathering evidences of the *what* rather than the *why*, though inevitably, as we shall see, the vast and overwhelming amount of data it accumulated led to some of its members trying to work out a comprehensive theory of the causes of these cycles, of the *why*.

In any case, this was Hoover's lasting legacy — a profoundly positive though unsung one — from the Great Depression and his otherwise blighted presidency, for it led in 1947 to the first "comprehensive" publication of what the Foundation had discovered. This was a book co-authored by Dewey and

---

2 "Edward R. Dewey," Wikipedia, en.wikipedia.org/wiki/Foundation _for _the_Study_of_ Cycles, p. 1.
3 Ibid.
4 foundationforthestudyofcycles.org p. 1, emphasis added.

his associate, Edwin F. Dakin, called *Cycles: The Science of Prediction.*[5] It is by means of this book that we shall gain our initial entrance into the deep physics connection between cycles of human behavior, economics, and energy technologies.

A. An Overview of Dewey's Database

Anyone reading Dewey's and Dakin's *Cycles: The Science of Prediction* will immediately be struck by the sheer amount of data — which is but a tiny *fraction* of the data gathered by the Foundation in its decades of research! — that is crammed into its comparatively short 255 overstuffed pages. There are graphs of the growth curves of pumpkins, yeast cells, and fruit flies,[6] of whole countries like Sweden, the United States,[7] France, Algeria,[8] of populations, of manufactures,[9] of exports and imports,[10] of the growth of railroads, shipbuilding, and automobile building industries,[11] of cattle, corn, cotton, wool, wheat, malt liquor, lumber, and cotton spindle production, coal, copper and lead production,[12] of airplane and natural gas production,[13] of paper and wood pulp production.[14] There are charts of the rises and fall of Atlantic salmon populations,[15] of lynx and tent caterpillar population cycles,[16] even of the cycle of ozone levels at London and Paris.[17]

There are charts of 54-year periodic cycles in wholesale prices, wages, and interest-bearing securities, of coal production in England and coal consumption in France, of pig iron and lead production in England.[18] There are charts and graphs of nine-year cycles of prices in manufactured goods, common stocks, wholesale prices.[19] There are three-and-a-half-year cycles of industrial production and common stock prices,[20] 18-year cycles of real estate activity, housing and building construction,[21] marriages, wheat acreage,

---

5   Edward R. Dewey and Edwin F. Dakin, *Cycles: The Science of Prediction* (New York: Henry Holt and Company, 1947. Reprint by Kessinger Publishing). ISBM 1436710219.
6   Ibid., pp. 12–13.
7   Ibid., pp. 14–15.
8   Ibid., pp. 16–17.
9   Ibid., pp. 24–25.
10   Ibid., pp. 26–27.
11   Ibid., pp. 32–33.
12   Ibid., pp. 36–41.
13   Ibid., pp. 44–45.
14   Ibid., pp. 46–47.
15   Ibid., p. 53.
16   Ibid., pp. 54–55.
17   Ibid., p. 56.
18   Ibid., pp. 80–81. For the discussion of the phenomenon of the 54-year cycle, see pp. 69–86.
19   Ibid., pp. 89–90.
20   Ibid., pp. 106–107, 110–111.
21   Ibid., pp. 118–119.

bricks,[22] of construction in Hamburg, Germany, of skyscrapers in New York City and Chicago,[23] loans and discounts and railroad stock prices,[24] and on and on it goes.[25]

What emerged from all this vast accumulation of data was that the growth curve of things — no matter what they were, from yeast cultures to countries — was *the same,* and that there were *discernible wave forms, cycles, of varying years' duration,* in all manner of things, from stock prices to ozone levels to atmospheric electricity, and that these cycles were not only discernible but *quantifiable* because they were *regular*. And because they were *regular,* they were not only *inevitable* but also *predictable.* As we shall see, in this inevitability and predictability, there lies quite a tale, for that story is in fact what this book is about.

But needless to say, the possession of such a database, not to mention its potential in the hands of those who knew about it, made it a "tool to possess" for those who stood to benefit financially and politically from such knowledge, and this in turn may account for the fact that, while a public foundation, few people actually *know* about it, and fewer still have an inkling of what its vast database implies about the structure of the physical world, and about those who would, through technological means, manipulate that structure for whatever purpose, and, in this regard, one need only recall that Dr. David Li's Gaussian copula formula was precisely a technique of such manipulation. But we get some measure, perhaps, of the Foundation's true importance and the true significance of its research by the presence within its membership and lists of published papers of at least one member of a prominent international bank. For example, in its library of available papers on its internet website, one finds the following interesting catalogue of papers, and these, again, are but a few of those publicly available. There are many more available to members of the Foundation:

1) A paper by Theodor Landscheidt, Director of the Schroeter Institute for Research in Cycles of Solar Activity, Nova Scotia, entitled "Cosmic Regulation of Cycles in Nature and the Economy." The short abstract of this paper is also illuminating, for the paper is

---

22   Ibid., pp. 120–121.
23   Ibid., p. 112–123.
24   Ibid., pp. 130–131.
25   On its website the Foundation for the Study of Cycles states that it is "dedicated to the interdisciplinary study of finding and analyzing recurring patterns. This includes the economy, natural and social sciences, the arts and more as *more than 4,300 natural cycles have been documented with interrelated patterns.*" Concerning its mission it states that "Our Mission is to discover, understand *and explain the true nature and origin of cycles, thereby solving the mystery of recurrent rhythmic phenomena,* as has been observed in both the natural and social sciences." (Emphases added) By 2006, it claimed more than "2,500 members in 42 countries." (www.foundationforthestudyofcycles.org)

about "cycles in nature that can be understood and predicted, and that are connected with human behavior, especially in the economy."[26] Note that what is implied here is a direct relationship between solar cycles and human economic activity.

2) A paper by Ray Tomes, Director, Advanced Management Systems, Ltd., in New Zealand, entitled "Towards a Unified Theory of Cycles." The short abstract of this paper is equally illuminating, for it states that "Evidence suggests that earth-based cycles are caused by lunar and solar effects on earth's weather system. New ideas are presented to explain both *the mechanism by which planetary alignments cause solar variations,* and the reasons for detailed harmonic relationships."[27] This is a bombshell, and will play a major role throughout this book, for observe what Tomes has just stated, and stated very explicitly: planetary alignments constitute *some sort* of causal mechanism on the cycles of solar activity, and with that, we are but one step away from astrology. There is but one problem with that conclusion, and that is that, with the Foundation, one is dealing with a group that has massive amounts of empirical, quantifiable, and regular cyclic data on which they may base their theoretical conclusions. And one need only compare this massive *historical* database, and the presence of at least one known member of the international banking community, with Dr. David Li's Gaussian copula formula which deliberately set aside historical data, to perceive the profound difference between the two approaches, and to perceive the implication that, at some point, someone "in the know" must have known what the real historical cycle would be.

3) Finally, and for our immediate purpose, there is a paper by Mike Niemara who is stated simply to be an "Economist, Mitsubishi Bank, New York," entitled "Forecasting Turning Points in the International Business Cycle." While the title of this paper is what one might expect from an economist working for a very large Japanese international bank, the abstract, again, is not, for besides referring to the use of "new statistical techniques" for the "monitoring and forecasting of turning points in the international business cycle," the abstract ends in a most unconventional fashion by stating that the paper will also deal with "the role of sunspots in economic activity"![28]

---

26 www.cycleslibrary.org/synchronies/pg_0004.htm
27 Ibid., emphasis added.
28 www.cycleslibrary.org/synchronies/pg_0004.htm

From this one may reasonably conclude that at least *some* in the international banking community have perceived a *physics* connection to macro-economic activity and cycles, and are bent on discovering what it is. One may envision this as a kind of "econphysics," for if there *is* a physics connection, then it stands to reason that if one has a "standard public consumption" physics view of physical systems as closed systems, then one will perforce have a similar view of economics. But if, as these three examples of the Foundation's published papers attest, both physics and economics are based on interrelated and hence open systems, then one will have a corresponding economics and physics.[29]

B. The Inevitability and Predictability of Cycles: Closed vs. Open Systems

But what did Dewey and Dakin themselves state about these cycles and their implications and significance?

In the introduction to their book, Dewey and Dakin state their case with careful circumspection:

> There are those who, admitting that economics has not been an exact science, also insist that it cannot be, in the sense of predicting outcomes in human affairs. There are even some who consider prediction regarding human life as a kind of impiety — or fakery, at best.
>
> .....
>
> The reader will be introduced to a method of thinking about the future which — new though it may be to him — seems definitely to have proved of value. It is this method which is of fundamental importance — an importance greater than any specific conclusions to which it may lead. For on its validity depends the whole value of the conclusions.

---

29   It is also worth pausing to mention another interesting implication of the economist of Mitsubishi being present as a Foundation member. Search as I did for the presence of similar members from large international Anglo-American or European banks, I could find none. This does not mean there aren't any, but that they may be members whose papers and membership in the Foundation are not published or known. But the presence of a Mitsubishi economist is interesting in its own right, for Oriental societies are well-known to have, from their very earliest cultural and religious origins, essentially *cyclic* views of history and hence of human social activity. The presence suggests, then, that they are intent upon putting empirical scientific and quantifiable data in the service of recovering the real scientific bases of these ancient cosmological views. In short, they are doing what I called paleophysics in my very first book, *The Giza Death Star*. (Q.v. *The Giza Death Star,* chapter three. The paleophysical approach to ancient texts and myths was further developed in *The Giza Death Star Destroyed*, chapters one through three, and in *The Cosmic War: Interplanetary Warfare, Modern Physics, and Ancient Texts,* chapters one through nine. This present book is intended to extend this analysis over a broader range of data.)

....
> It is hoped that the reader's reward will be the discovery that in economics, as in other sciences, we are apparently dealing with laws regarding rhythmic human response to certain stimuli that give *a remarkable working tool to any man who is responsibly concerned with future outcomes — whether he be businessman, community leader, or statesman.* [30]

Of course, the idea of a "remarkable working tool" in the hands of those "responsibly concerned with future outcomes" implies the same remarkable tool could be used by those "*irresponsibly* concerned with future outcomes" as well. In short, behind their carefully measured prose lies the true significance of their work: it is an extraordinarily powerful tool of prediction, and, to that extent, of manipulation.

This they admit a little further on, when they hint at the potential physics basis of their work:

> The discovery that the law of averages applies to humanity — that certain activities of people, viewed en masse, fall into definite patterns, some of which repeat themselves with periodic rhythm — promises to be of great aid in making economics function as a true science.[31]

This appeal to a "statistical approach" to human behavior *in the aggregate* reveals the potential physics implications, for a similar such approach is used in quantum mechanics, where the random motions and positions of individual particles is never known, but a statistical picture can emerge of their behavior in the aggregate. As will be seen, this analogy to physics only grows, and becomes more explicit, as we proceed.[32]

But then, shortly after implying this "statistical physics" approach to sciences of human behavior, Dewey and Dakin make a statement pregnant with implications:

> When a people finds that predictions of many financial advisers, statesmen, historians, and other proclaimed experts are seldom better than the predictions of the astrologers, our social sciences have demonstrably not been earning their way. It is time for action.[33]

---

30 Dewey and Dakin, *Cycles: The Science of Prediction,* p. xii, emphasis added.
31 Dewey and Dakin, *Cycles: The Science of Prediction,* p. xiii.
32 It is worth noting that this connection between physics and the modeling of social behavior in the aggregate was noticed by the celebrated and famous physicist and science fiction author, Isaac Asimov, in his well-known (and perhaps not just coincidentally-named) *Foundation* trilogy.
33 Ibid., p. xiv.

Obviously, from the tenor of their remarks, Dewey and Dakin do not repose much scientific basis or confidence in astrology. Yet, by appealing to it as an illustration of the state of the social sciences, including economics, they have implied a connection between the two. It is a connection that, as we shall see, only looms larger and larger as our examination of their book *Cycles: The Science of Prediction* proceeds. Indeed, they can hardly have been oblivious to the connection, and may have been trying to head off any criticism from certain quarters that they were indeed proposing a sophisticated kind of astrology.

Having stated this peculiar comparison and caveat, they then proceed to distinguish between a "theoretical economics" and an "empirical economics," again implying yet another analogy to physics:

> This book is an attempt to show, in an elementary way for the reader unfamiliar with this form of research, how some of the inept arguments over economic outlooks can be avoided by using a few facts that should now be familiar to all. This application of a new method to a study of economic activity, while relatively young, seems nevertheless more promising, in offering results, than traditional economic theories that fill textbooks with opinions and arguments over whether a given cause is really an effect, and vice versa.[34]

In other words, one may imagine a "theoretical economist" filling blackboards with complicated models and differential and integral equations, and on that basis, making analyses of "causes" and of "effects" with no real practical utility or basis in data — Dr. Li comes to mind once again — or at best, a basis in a rather narrow selection of data. In fact, some data may fly completely in the face of the proposed theoretical model. In this, the analogy to much of modern theoretical physics is very striking, where a certain class or type of data is ignored or swept under the rug because it is either contradictory to the "known" theory, or because the accepted and reigning theory is simply unable to account for it, and because there may not be — by such "theoretical physicists'" lights — any way to account for the data by any theory known to them. This, for Dewey and Dakin, was exactly the state of modeling in standard economics in their day. For them, they state their case with a classical "empirical scientist's" concision: "(The) 'forecasts' are written by the data.... They do not rest on the opinion of any man, or men. They are, in effect, the *probabilities* of tomorrow."[35]

But what exactly did all this vast mountain of data — from cycles of

---

34   Dewey and Dakin, *Cycles: The Science of Prediction*, p. xv.
35   Ibid., emphasis in the original.

atmospheric electricity to tent caterpillar populations — indicate?

A glance at a few of Dewey's and Dakin's charts of growth patterns in various things will say more than any words possibly could.

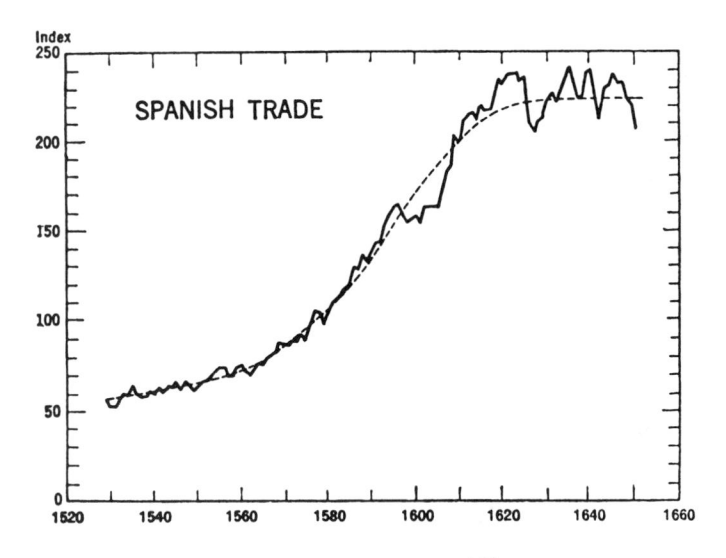

Graph of Spanish Trade[36]

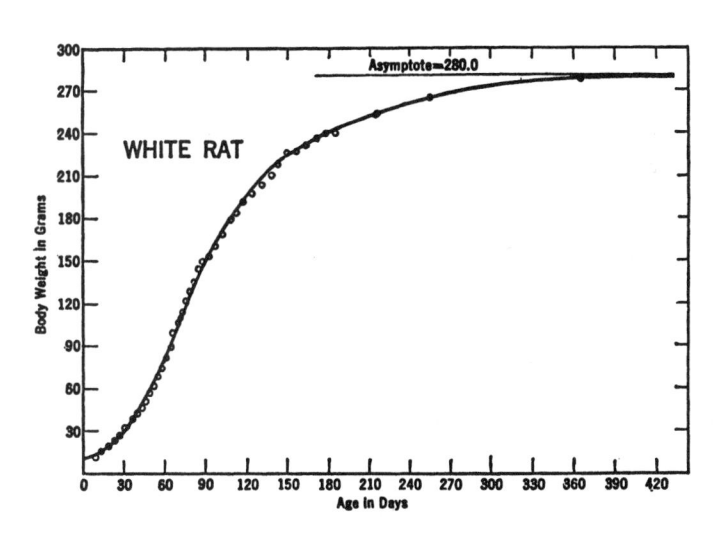

Growth of White Rat Population[37]

---

36 Dewey and Dakin, *Cycles: The Science of Prediction,* p. 9.
37 Dewey and Dakin, *Cycles: The Science of Prediction,* p. 11.

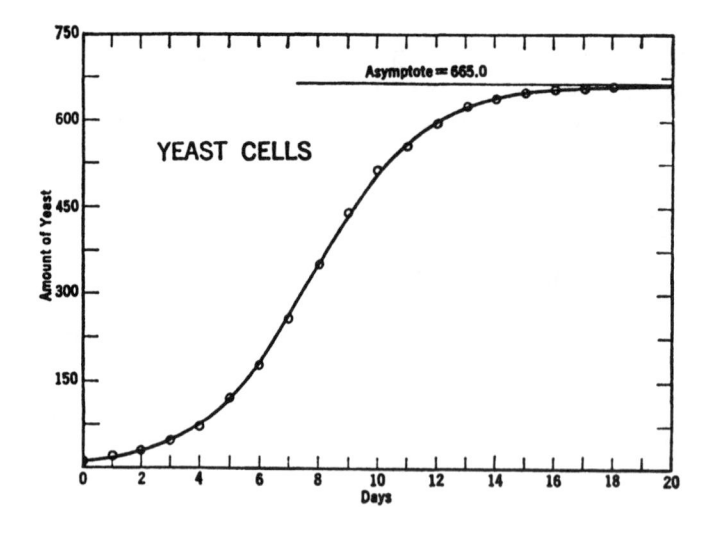

Graph of the Growth of Yeast Cells[38]

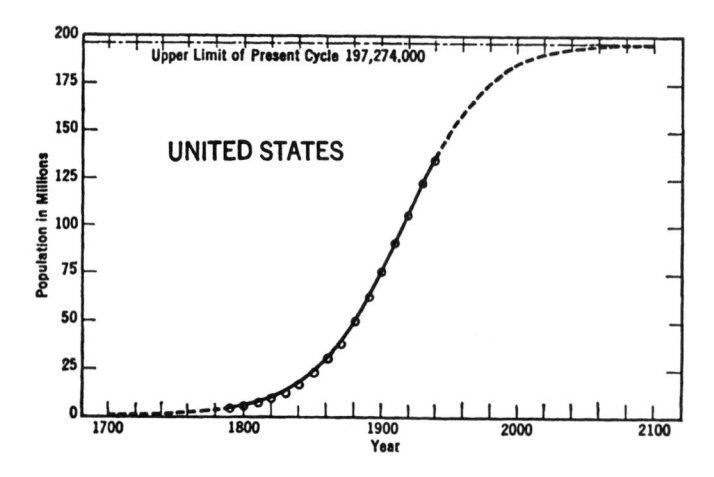

United States Population Growth[39]

Dewey and Dakin explain the meaning of these and similar graphs with simplicity and concision:

> The facts of growth are common knowledge to most mothers, who are encouraged by doctors to keep a weight chart around the nursery, for reference at weighing time. All healthy babies, like other healthy organisms, show large initial rates of growth — over 100 per cent for

---

38 Ibid., p. 13.
39 Ibid., p. 15.

babies the first year. As they get older, the rate of growth gradually falls off. At the approach of maturity, the rate growth finally reaches zero.

Why it is that an organism stops growing we do not really know....

But knowledge that such an inhibiting factor does exist is important to us, even when we cannot explain it. We find it reasonable that such a factor should be at work in organisms like a baby or a tree, just because we are used to observing it in action. But it also works in other kinds of organisms, such as human institutions and business organizations.[40]

In other words, the pattern of growth, when appropriately graphed, is almost universal and identical, regardless of the time frame in which such growth is measured, whether that of mere minutes, days, or whole decades.

And this necessitates a comment about Dr. Li's copula formula, and indeed, about the whole nature of privately-created debt money, versus the actual pattern of economic growth. Under Li's formula, and indeed, behind the whole principle of money-as-debt creation, there lies one very obvious fact, namely, that there is no limit on the number of credit default swaps, or on the amount of debt, that can be created. Indeed, as the law of compound interest shows, such debt only grows exponentially, like a cancer. But as Dewey's and Dakin's endless examples show, *actual growth within an economic system reaches a peak and begins to level out, barring the introduction of new factors such as a new technology, into the system to open it up once again.* Thus, under such considerations, the interest on debt, or even on debt swaps, can never be repaid.

For example, when examining such cyclic graphs for industrial production in various key industries in the United States, as well as for its overall population growth and general economic trend, Dewey and Dakin came to what was then, in economically booming post-World War II America, a rather disturbing and radical conclusion:

> ...(The) trends existing in a number of our great industries show definitely that we have been reaching a period of basic "maturity" in our whole economic development. This is a fact of enormous implications that reach in many directions. The implications are so great, indeed, that many people (as usual with humanity) find it easier to deny the fact, in heated argument, than face it honestly and then proceed to deal with it.[41]

---

40   Dewey and Dakin, *Cycles: The Science of Prediction*, p. 1.
41   Ibid., p. 5.

To put it succinctly, the United States was reaching the upper limit of growth in almost all crucial categories of activity. And, as their own graph of U.S. population growth that we have reproduced previously itself indicated, this limit would begin to be reached around the years 2000 to 2100. It was a cycle reflected in other key areas.

1. Criticism from the Conventional

The analogy between physics and economics in their theoretical and practical pursuits highlights a series of criticisms conventionally-minded economists such as Dr. Milton Friedman have leveled against Dewey's and Dakin's *Cycles*. According to Friedman, their book

> is not a scientific book; the evidence underlying the stated conclusions is not presented in full; data graphed are not identified so that someone else could reproduce them; the techniques employed are nowhere described in detail.... Its closest analogue is the modern high-power advertisement — here of book length and designed to sell an esoteric and supposedly scientific product. Like most modern advertising, the book seeks to sell its product by making exaggerated claims for it... showing it in association with other valued objects which really don't have anything to do with it... keeping discreetly silent about its defeats or mentioning them in only the vaguest form... and citing authorities who think highly of the product.[42]

But Friedman, too, later in life succumbed to the temptation to write a book about economic science for a popular audience called *Free to Choose,* one of the books influential in the popular culture of the Reagan era, and his own earlier criticisms against Dewey and Dakin might justifiably be leveled at his own attempts at popularization in that book.

More detailed criticisms came from economist Murray Rothbard, and with these criticisms, the analogy between physics and economics becomes even more acute:

> Any such "multicyclic" approach must be set down as a mystical adoption of the fallacy of conceptual realism. There is no reality or meaning to the allegedly independent sets of "cycles." The market is one interdependent unit, and the more developed it is, the greater the

---

[42] Milton Friedman and Max Sasuly, "Review of *Cycles: The Science of Prediction,*" *Journal of the American Statistical Association* (March 1948, 43 (241): 139–145, cited in "Edward R. Dewey," *Wikipedia,* en.wikipedia.org/wiki/Foundation_for_the_Study_of_Cycles, p. 2.

interrelations among market elements. It is therefore impossible for several or numerous independent cycles to coexist as self-contained units. It is precisely the characteristic of a business cycle that it permeates all market activities.[43]

But Rothbard's critique misses the mark, for nowhere do Dewey and Dakin claim that their numerous cycles are "independent" of each other. All they claim is that, whatever phenomenon as one chooses to examine that endures over time, that phenomenon will demonstrate a cycle or wave pattern. They do not claim for them a self-contained independent *closed system* existence free from interrelationship with other such cycles. Indeed, as will be seen in the next section, they argue the exact *opposite* of what Rothbard says they do.

The physics analogy of Rothbard's critique is thus evident, for by claiming that their multicyclic approach "must be set down as a mystical adoption of the fallacy of conceptual realism," he is suggesting a kind of "Copenhagen" school of economics analogous to that in quantum mechanics, for whom the equations of quantum mechanics did *not* refer to any underlying physical reality as their basis, but rather such equations were solely mathematical artifacts of the theory.[44] Similarly, for Rothbard, "there is no reality or meaning to the allegedly independent cycles" *because* they are interrelated. For Dewey and Dakin, they *are* real precisely *because they are open systems and interrelated.* The bottom line then is that Rothbard's critique is an ideological and philosophical one. The problem, however, is that both Dewey and Dakin, and their critics, did not realize the true nature of what they had discovered, namely, that they had discovered something less about economics than about *physics,* and that the economic activity their cycles were tracking were but a subset of a larger cyclic class of phenomena whose true nature and characteristics were the result of a very deep physical mechanics of the physical medium itself. But that insight must wait for future chapters. We have already seen, however, that Dewey and Dakin clearly must have suspected as much, since they included cyclic data about purely physical phenomena, and, as we have also already seen, many members of the Foundation have presented papers linking the two in a kind of "econophysics," suggesting a purely physics-based influence on the patterns of cyclic human activity. This was a conception that, once again, recalls astrology, and Dewey's and Dakin's own remarks about economics and astrology. Such remarks suggest that they may have begun to suspect some deep and profound relationship between their economic cycles and very

---

43 Murray Rothbard, *Man, Economy, and State* (Nash Publishing Co., 1961), cited in "Edward R. Dewey," *Wikipedia,* en.wikipedia.org/wiki/Foundation_for_the_Study_of_Cycles, p. 4.

44 Q.v. the discussion in my *The Giza Death Star* (Adventures Unlimited Press, 2002), pp. 126-134.

ancient conceptions of physical influences upon mankind epitomized by that ancient craft.

C. Cycles, Trends, and Cycles Upon Cycles

So exactly how *do* Dewey and Dakin demonstrate their awareness that all the vast number of cycles their Foundation has tracked are indeed interdependent? The answer to this question lies (1) in their distinction between cycles and trends, (2) in their conception of a periodic cycle, and (3) in what they call "overlays" of cycles. With these "overlays," as we shall see, there is an exact physics analogue, and in that analogue, profound implications.

1. The Period Patterns

Cycles may easily be distinguished from mere trends within an overall periodic rhythm by a glance at one of their many graphs. Note on the following graph how the basic waveform of the cyclic period is graphed, versus the actual measured data, some of which falls above, and some of which falls below, the theoretical value of the wave form itself. Those smaller peaks and troughs within the overall pattern they designate as "trends," as distinct from the overall periodic cycle itself. If one examines the chart carefully, it is evident that the psychological effects of a "trend" will vary depending on where in a cycle it occurs. For example: if a downward trend occurs on the upward curve of a cycle, the effect will not be as severe. If it occurs on a downward portion of a cycle, it will be felt as correspondingly more severe.[45]

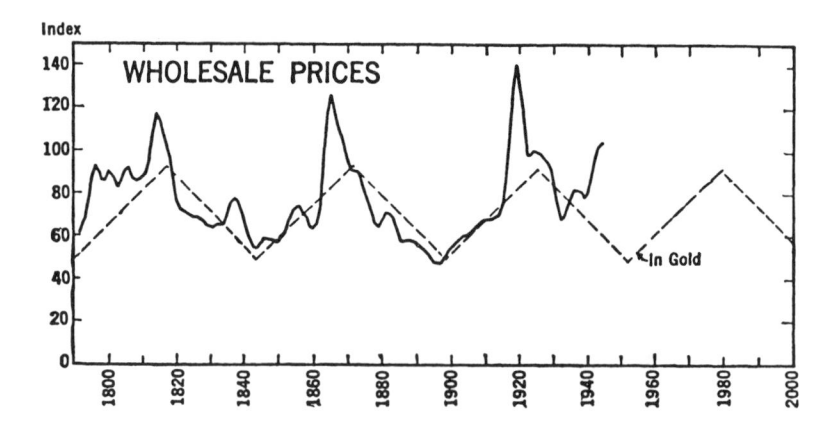

54-Year Periodic Cycle of Wholesale Prices[46]

---

45   Q.v. Dewey and Dakin, *Cycles: The Science of Prediction*, p. 50.
46   Ibid., p. 70.

As we shall discover, this distinction between a mere trend and an overall cycle will have profound implications in their understanding of how much, or little, human actions, including political action, can exacerbate or alleviate an overall cycle.

Their remarks in this respect are illuminating:

> We shall quickly discover that the so-called business or economic cycle is in reality a composite of many different cycles. We shall be particularly interested in studying them for evidences of *rhythm*. For if and when we find rhythm we can, *to the extent justifiable,* predict recurrence; and we shall have predictability on the basis of much more than finite logic and guess.
>
> It has long been assumed that if we could only isolate the *causes* of economic cycles we could then effectually prevent them, and so eliminate the downsweeps that plague highly organized societies.
>
> Many volumes have been written to discuss the cases. **Here we shall be far less concerned with the cause than with the timing.** [47]

Note that Dewey and Dakin explicitly state that the "economic cycle is in reality a composite of many different cycles," and thus expose the criticism of Murray Rothbard previously cited — that they do not take into consideration the complexity of the interrelationships of components of an economy — as utterly unrelated to anything Dewey or Dakin actually wrote! Their real "heresy," as we shall see, lies not in the interrelationship of economic cycles, but in proposing that there may be even deeper relationships between economic cycles and cycles of purely *physical* phenomena.

In any case, it is worth noting what they state about rhythms or rhythmic cycles and periods:

> *Rhythm,* or *rhythmic cycle,* will ordinarily be used to denote a cycle which repeats itself at rather uniform time intervals. Rhythm implies a kind of beat. Thus the beating of the heart is rhythmic, though it may be slightly irregular.
>
> *Perdiocity, periodic cycle,* or *regular cycle,* will be used to denote a cycle which repeats itself **at mathematically exact intervals**. True periodicity, like a true straight line or a true circle, does not exist in nature; but we have close approximations.[48]

---

47 Ibid., p. 50, boldface emphasis added, italicized emphasis in the original.
48 Dewey and Dakin, *Cycles: The Science of Prediction,* p. 51, boldface emphasis added.

Here is yet another link, and a potential clue, to the deeper physics that may be involved in such interrelationships, for such cycles are precisely quantizable, and these quantities, these measurable cycles, may be giving out profound clues about the nature and structure of space-time itself. Dewey and Dakin were not oblivious to this physics connection and implication, but on the contrary, carefully selected a specific physics analogue to explain their conception of deeply interrelated economic (and other) cycles. To see what this physics analogue is, one must examine what they mean by interrelated cycles.

2. "Overlays": Combined Cycles as Multiwave Modulation, and Possibly as Interferometry?

Briefly put, interrelated cycles are really nothing but super-positions of the wave form of one cycle upon that of one or more others. For example, among the many quantifiable types of cycles operative in various realms of activity — whether human or otherwise — Dewey and Dakin devote a significant portion of their book to the overview of a nine-year cycle,[49] a nine and two-thirds year cycle,[50] a four-year cycle,[51] a 41-month solar cycle and other types of solar cycles,[52] a 35-year weather cycle,[53] a 22-year weather-solar cycle,[54] and of course the 54-year cycle whose graph we have previously cited.[55]

As indicated, Dewey and Dakin, no less than their more conventionally-minded economist critics, view economic cycles as a complex system of interrelationships, but unlike their critics, do not despair of quantifying those relationships, not in terms of their *causes and effects* but in terms of their *beginning and periodicity in time*. Thus, the interrelationship of all these cycles expresses itself in an exact model: that of the superposition of the wave forms of the cycles themselves, or what they call "overlays."

---

49 Q.v. Dewey and Dakin, *Cycles: The Science of Prediction*, 56ff.
50 Ibid., p. 54ff.
51 Ibid., p. 57f.
52 Ibid., pp. 57–59.
53 Ibid., p. 60.
54 Ibid.
55 Ibid., p. 70. It is interesting to note that Dewey's and Dakin's 54-year wholesale price cycle last dipped ca. 1951–54, and then began the next 54-year period of rising prices, which would put its peak ca. 2005–2008, precisely during the current "derivatives crisis" and its initial aftermath, a sharp decline in prices.

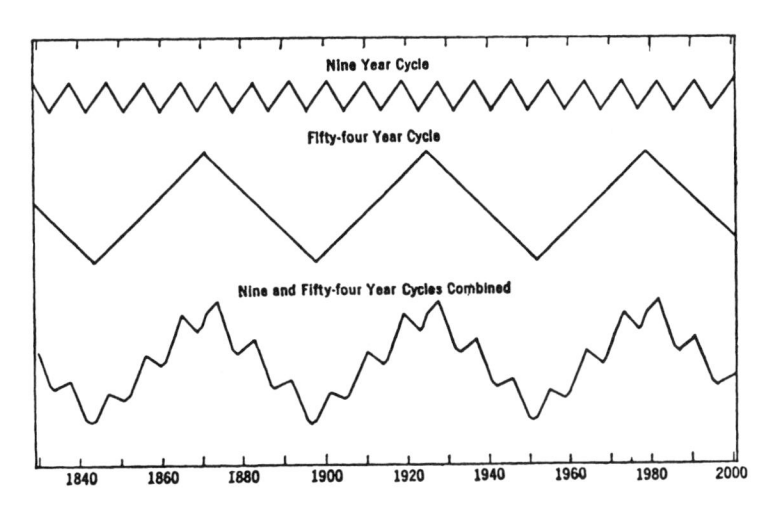

Dewey and Dakin's Example of the Overlay of the Nine-Year and 54-Year Cycles [56]

At the top of this graph is a simplified version of the nine-year cycle. In the middle is the 54-year cycle, and at the bottom is the "overlay" or composite of both cycles. It should be noted that the longer and lower-frequency 54-year cycle retains a "dominant characteristic" in the composite over its shorter and higher-frequency nine-year cycle. This too, as we shall discover, is yet another important clue to a possible deeper underlying physics.

According to Dewey and Dakin, one of the first to notice the possibility of a connection between economic activity and physics was Sir William H. Beveridge, who

> Found several true periods that gave strong evidence of correlation with known weather rhythms and sunspot phenomena. And a T= 54.00 years (which is without such a correlation) he also found a relatively significant concentration of energy.[57]

Of course, this much was, in a certain sense, obvious, for solar activity does affect terrestrial weather, and terrestrial weather can in turn affect agricultural production, which in turn can affect overall economic activity. There does not seem to be much here by way of a deeper physics.

But it was when certain cycles — particularly the nine-, nine and two-thirds, and 54-year cycles — were overlaid that one began to notice something very significant:

---

56   Ibid., p. 96.
57   Dewey and Dakin, *Cycles: The Science of Prediction*, p. 73.

As Schumpeter has said:

> "It is clear that the coincidence at any time of corresponding phrases of all three cycles will always produce phenomena of unusual intensity, especially if the phases that coincide are those of prosperity or depression. The three deepest and longest "depressions" within the epoch covered by our material — 1925–1930, 1973–1878, and 1929–1934 — all display that characteristic.[58]

As will eventually be seen, the overlay of these cycles and their concurrence at periods of various "great depressions" in American economic history do indeed point to a deeper physics, but one in which, however, the astrological connection is even more in evidence.

Needless to say, however, this approach of "overlays" or superpositions of various economic and business cycles — not to mention their combinations with other cyclic rhythms having at first glance nothing to do with economics — throws the normal "cause and effect" analysis of conventional economic thinking and theory into a cocked hat.

> Most of us, taught to think in conventional terms of cause and effect, and trained in a system where the educational wares are often cut to the lowest common denominator of mass intelligence, find it hard to remember that the human mind is very finite. It can never hope to grasp in any situation all the causes that may work together to produce any given result. That is why judgments are so faulty when we reason solely on the ordinary cause-and-effect basis. Some if not most of the actual causes will escape our knowledge, and those we do take into calculation may be pure assumption. If despite this we reach a correct conclusion, it is a triumph of accident — or of intuition.[59]

For Dewey and Dakin, the fallacy of such cause-effect thinking in economic cycles is aptly illustrated by the similarity of cyclic activity in real estate cycles and marriages within the United States. This similarity

> Could readily lead to some false assumptions. It would be simple enough to reason that people who get married start to think about house-hunting and home-building, and that this results in building projects. Or one could start his reasoning in reverse, and end in the same place, by saying that building stimulates prosperity, and when people feel prosperous they get married, and when married they

---

58 Ibid., pp. 90–91, citing Schumpeter, *Business Cycles* (McGraw-Hill, 1939), p. 173.
59 Dewey and Dakin, *Cycles: The Science of Prediction*, p. 114.

buy the homes that contractors have previously erected to attract customers who feel prosperous because money is being spent to erect buildings for people who will get married.

Such reason-why arguments chase their tails. Nor is this statement of them unusual, except in brevity. Whole books have been written that incorporate such "reasoning" into learned syllables. We should be prepared to ignore reasoning of that kind, but we are justified in asking *why* our building statistics should formulate themselves in such exceedingly regular rhythms. Before research of the future supplied us with a definitive answer to this particular query, we may hazard a postulate that man's mating instinct and man's building instinct may be aboriginally associated in his being, just as they extend down the biological scale to the birds. If both instincts proved to flow in the race in a common rhythm, it should not seem too surprising. Perhaps economists could well join forces with biologists in the new kind of research such a problem suggests.[60]

Note then, that far from viewing each cycle in their "overlays" as independent realities as their critics such as Rothbard would suggest, not only do Dewey and Dakin view the entirety of economic activity as a complex overlay of such cycles, but they level almost the same criticism at their detractors, in that by focusing too narrowly on purely "economic" activity and cycles, their critics may in fact be ignoring cyclic clues from other disciplines that may afford genuine insight as to the real natures of their causes.

Thus, they call for an interdisciplinary approach. They also observe that, as comparison between disparate and discrete disciplines grows, so too does the suspicion that the discrete phenomena are related in some deep fashion, that they are all clues to a deep connection in the physics of all observed cycles, no matter what their subject matter:

> In the various categories of rhythms we know, some are as familiar to the man in the street as day and night and the tides; others are relatively unknown to him. Analytical work on these rhythms has been developing in what might be called the interstices of the sciences. We have as yet only the beginnings of a science of rhythms *per se.* But in each field of science some group of workers becomes especially concerned with the subject of rhythms in the special field of phenomena he is observing; such a group eventually compares notes with a corresponding group engaged in some other scientific field; and

---

60   Dewey and Dakin, *Cycles: The Science of Prediction*, pp. 116–117, emphasis in the original.

slowly all concerned begin to suspect that they are dealing, however diverse the fields, with phenomena somehow related.[61]

As is by now apparent, Dewey and his Foundation for the Study of Cycles had long since suspected such deep connections between disparate and discrete cycles, for it was founded precisely as a means of coordinating data from all fields of inquiry in an interdisciplinary fashion.

3. Inevitability, Predictability, and Human Actions

There is another implication to Dewey's and Dakin's analysis, and indeed, to the work of the Foundation for the Study of Cycles as a whole. It is an implication already encountered, but it is best to restate it here: if cycles — particularly economic cycles — are of regular and quantifiable periodicity, then this implies that they are *inevitable, and because inevitable, predictable.* This in turn would seem to limit the scope of human action and its ability to influence such cycles merely to affecting *trends*, rather than the cycles themselves. Indeed, citing Schumpeter's *Business Cycles* once again, they note that when one studies the various economic policies of England, the United States, and Nazi Germany in response to the Great Depression, they noted that of all the various measures each country took, including Hjalmar Schacht's subsidized exports and various systems of marks in Germany,[62] to Roosevelt's New Deal programs in the U.S.A., the rises in prosperity in each country nevertheless "arrived strictly as might have been anticipated by anyone projecting the familiar economic waves of the past."[63] In other words, whatever the *trends* the respective governmental policies in each nation might have induced, the overall cycle itself was destined to turn precisely when it did, regardless of governmental policy or political action. Financial policy could *amplify or damp* the overall trend within a cycle, but could never arrest or reverse the cycle itself. And the reason is again clear: the cycle itself was evidence of a deep physics being played out on the stage of aggregate human activity.

Indeed, such policies or actions oftentimes function merely as devices of concealment, cloaking the reality of the inevitability of such cycles by appearing to manipulate the cycle itself by measures that, in fact, manipulate only the trend:

> Devices like the Bretton Woods Economic agreement, or Treasury grants made as political "loans" to foreign nations, may tend for

---

61 Dewey and Dakin, *Cycles: The Science of Prediction,* p. 140.
62 Ibid., p. 78.
63 Ibid., p. 79.

temporary periods to mask this truism — much as the working of the Federal Reserve Act tends to conceal from the general public the fact that we print dollars to meet government deficits. But our chart says forthrightly, in statistical language, that on the basis of the long-established trend, and in terms of foreign trade handled at a real profit, we should not be too optimistic in looking for such trade to be increased over the prewar volume.... All our chart can tell us is that, *failing a revolution of some sort,* the pattern as established will presumably prevail.[64]

Note then that, so long as prevailing conditions remain, the cycle itself is basically unaffected. It is when the system again becomes an *open system,* when changes of a radical nature, such as a technological revolution, occur, that the cycle itself can be directly affected.

4. Waves, Wars, and Revolutions

But barring this, humans inherently tend to resist the notion that economic cycles and waves may be beyond the ability of man to control. "They accept the need of adjustment to the weather, knowing it is out of man's control. But they resist passionately any suggestion that changes in the social and economic climate may be beyond the total rule of man's conscious will."[65] The only way that mankind could in effect change the inevitability of a given cycle of economic activity is precisely to introduce revolutionary change:

We now know enough about trend lines to realize that old ones merge into the take-off of new ones *only* when fundamental, even revolutionary, changes have occurred in the environment and its organization, and perpetuate themselves....

Remember Pearl's (fruit fly) bottle. Imagine for a moment that we Americans are the (fruit flies). We are reaching the upper asymptote of our (population growth) curve. The invention of a few new gadgets by industry — indeed, the invention of a whole new industry itself — is a force in no way adequate to change the relationships in our bottle. *We must have a whole new bottle.*[66]

In other words, we now find the economic inevitability behind the strategy outlined in chapter one; the global elite, in order to maintain any sort of

---

64 Dewey and Dakin, *Cycles: The Science of Prediction,* pp. 27–28, emphasis added.
65 Ibid., p. 191.
66 Dewey and Dakin, *Cycles: The Science of Prediction,* p. 222, emphasis added.

"sustainable growth" — to use one of their favorite catch phrases — must either expand humanity's "bottle" into outer space in a significant way, or perpetuate regional imbalances and conflicts and wars and the economic growth for some that this inevitably brings, *or* they must admit the growth of a revolutionary technology which completely changes the nature of the relationships inside the bottle itself. Failing this, they must export and then re-import manufacturing and technology from one region to another, and lower population, in a never-ending shell game, if the bottle-expanders of outer space and radically new technologies are not pursued. In short, the only real way to alter the inevitability of a cycle is to make the bottle an open system, and any open system invariably challenges the very basis of their power, unless, of course, the pursuit of revolutionary technology is one they can monopolize and employ to secure their hegemony.

5. Waves, Overlays, and Modulation: The Physics Analogues Employed by Dewey and Dakin, and Their Implications

Dewey and Dakin do much more than just *imply* a physics connection with all their allusions to closed and open systems, and to deep connections between economic activity and solar cycles and overlays of cycles. They also draw very explicit analogies to physics. We have noted that they draw upon cycles of various length — nine, four, 35, and 54 years to name but a few — but they are open to even more possibilities:

> Now, let us suppose that there are still longer waves in the universe — "Y" waves we may call them. Imagine some of the waves with peaks which come 3 ½ years apart, others with peaks 9 years apart, 18 ½ years apart, 54 years apart, and perhaps much farther spaced. It is not inconceivable that these longer waves could directly or indirectly affect the sun, weather, animals, and human beings, and that just as a red pencil may respond to light waves of only one length, so a particular organism might respond only to Y waves of one particular length.[67]

They are, in other words, proposing a kind of "econo-bio-physics," where particular kinds of organisms are transducers, are coupled harmonic oscillators, of waves of cycles of particular wavelengths, much like pipes in a pipe organ respond to different wavelengths and harmonics.

Invoking the analogy of sound, in fact, is not reading too much into their remarks, for they in fact do so by pointing out that the exact physical analogue of their "overlays" of cycles is that of sound waves being modulated together.

---

67 Dewey and Dakin, *Cycles: The Science of Prediction,* p. 158.

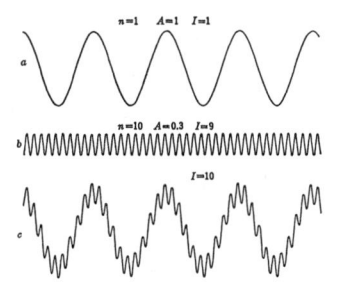

Overlay of Two Sound Waves[68]

Again, as with their overlay of the 54- and nine-year cycles, note that the overall characteristic of the lower-frequency, longer-wavelength wave is preserved even with the grafting of the shorter higher frequency sound wave on to it. This process is called modulation, and in this instance, the longer lower frequency wave has become a carrier wave. They extend the wave modulation analogy even further, by pointing our that any wave of a particular wavelength and shape will have an overtone, or harmonic series,[69] which are composed of *fractions* of that wavelength: one-half, one-third, one-fourth, one-fifth, and so on. They give the following chart of the graph of a sound wave from an organ pipe, and its first 12 overtones, or harmonics:

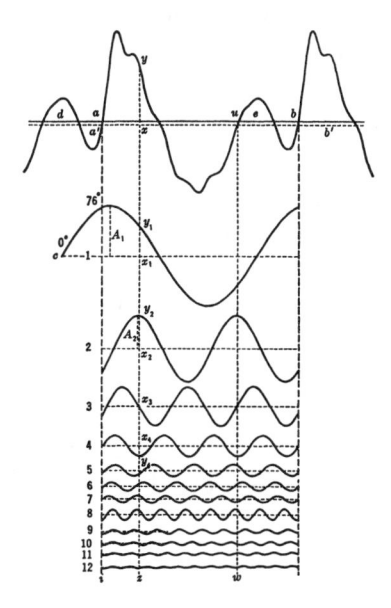

Harmonic Series of an Organ Pipe[70]

---

68   Ibid., p. 164.
69   Q.v. my *The Giza Death Star*, p. 113f.
70   Dewey and Dakin, *Cycles: The Science of Prediction*, p. 165.

Dewey and Dakin are strongly suggesting, then, that there may be a deeper physics of a *wave mechanics* involved in the overlay of *economic* cycles, and that correlations of sunspot activity with, for example, economic cycles of wholesale prices, may have a much deeper basis than merely the influence of the sun upon the weather, and of the weather on agricultural production and prices. In other words, they may *both* be immediate and manifest results of an underlying and more unified phenomenon.

But they go *much* further than even this, and in doing so, not only demonstrate that they are thinking in terms of a very deep, and indeed hyper-dimensional, physics, but also reveal some profound implications and questions that will preoccupy us throughout the remainder of this book. It is best to cite them at length in order to understand the full implications and significance of the deep physics connection that they are suggesting, and of its profound implications:

> To that end let us consider in terms of modern psychology and physics a few facts important for our approach to economic science.
>
> P.D. Ouspensky once asked his readers to make an experiment. Imagine, he said, that you live in two dimensions, instead of three. An easy way to do this is to imagine you are a being like a piece of paper, infinitely thin, living upon a table. You can look neither up nor down, for up and down are in a third dimension. You cannot even *think* up and down, or conceive it. For you have no thickness, and hence cannot even *imagine* thickness.
>
> Now in the center of this tabletop where you live, there is cut a slot. In this slot there revolves a wheel, so hung that half the wheel is always below the table, and half of it above. This wheel is solid and you can see only the edge of it. Let us imagine its edge is painted in four colored segments — black, white, blue, and red. As the wheel revolves, you observe it end-on, you of course do not know that it is a wheel you see. For you are a two-dimensional being, and therefore see only a *single line of color* along the tabletop. Occasionally, as the wheel slowly revolves, you do see that line change suddenly in color. Red will suddenly change to black, and black to white, white to blue, and blue to red again.
>
> Now, if you observe this phenomenon long enough, you will finally decide that when the red comes up, it will eventually cause black; and when the black appears, it will eventually cause blue. You will think you know the causes of the phenomena you observe.
>
> If a two-dimensional scientist is observing the phenomena, he will eventually discover a "law" in this continuity of event. Using this law,

he will be able to predict changes of color accurately. The scientist, by the use of mathematics, might also discover that a third dimension was necessary to account for the real phenomenon he saw in two dimensions only. But neither of you could *imagine* this third dimension as a sensory reality. Nor could you know the real nature of the causes operating there. The scientist would admit this frankly, saying his law merely described what happened, without explaining it. But you, untrained in such fine distinctions, would speak boldly of a "cause" being followed by an "effect." And each "effect" would in turn become a new cause (in your way of thinking) resulting in a further effect which followed. If you persisted in this belief, you might eventually resent being told that you knew nothing about the real causality.[71]

Clearly, Dewey and Dakin are suggesting that what we call cause and effect are really but artifacts of our "three-dimensionally-conditioned consciousness" grappling with phenomena that have their origins in a hyper-dimensional world of *more* than three dimensions.

One might therefore add to their analogy of a "hyper-dimensional" world *evidencing itself in cycles* that our scientist might, by dint of the same sort of mathematical techniques, deduce a very *different* object — the "wheel" — as the real cause and effect of that which we perceive as a cycle. And this brings them to a vital fact:

> What we call our recognition of *"cause"* and *"effect"* is *somehow associated with time,* and with our perception in time. This is important to understand, for what we call time is apparently only a mode of perception.... Ouspensky, approaching the problem from a psychological background, goes so far as to suggest that **time is the way we experience space in its higher dimensions. That is, the unknown dimensions of space are revealed to us in time.**[72]

In other words, not only are Dewey and Dakin suggesting that the sought-after deeper physics underlying their "harmonic overlays" and modulations of cycles might lie in a hyper-dimensional physics, but they are also suggesting that cause/effect thinking such as in the case of the sunspot-wholesale prices overlay cycle is in fact erroneous. It is not solar activity as the ultimate case,

---

71 Dewey and Dakin, *Cycles: The Science of Prediction,* pp. 192–193, referencing P.D. Ouspensky, *Tertium Organum* (Alfred A. Knopf, 1922), p. 65, emphasis added by me.

72 This profound insight suggests a connection to the views on time as a determinative of cause and effect relationships in a hyper-dimensional universe, that of the thought of Russian astrophysicist Dr. Nikolai Kozyrev. Q.v. my book *The Philosophers' Stone: Alchemy and the Secret Research for Exotic Matter* (Feral House, 2009), pp. 151–200.

with weather modification as the mediate cause, and wholesale agricultural price changes as the final effect, but rather, the solar cycle, the fluctuation of weather, the fluctuation of prices might all be the immediate, though admittedly interrelated, *three-dimensional effects of a hyper-dimensional cause.*

In such a model, our perceptions of them as "weather" and "solar activity" and even "price fluctuations" and so on, are similar to our perceptions of a beam of light that is split by a prism into its spectral components. In such a model, our three-dimensional reality is the prism refracting the several wave forms into its various constituents that we perceive as separate cycles. Moreover, Ouspensky, whom they cite, is quite clearly suggesting that time itself is not a dimensionless "scalar" entity, but rather an entity of multi-dimensional spaces which three-dimensional reality experiences *as a temporal flow*. Thus, their analogy of the modulation of waves upon a carrier wave, and of multiple harmonics of that wave, as in the pipe organ example, is meant to couple the notion of hyper-dimensionality to those harmonics of waves. This, as we shall see in the next chapter, is a profound clue.

D. Conclusions, Implications, and a Segue in the Form of Significant Questions

So what may be reasonably concluded about this cursory overview of the cycles research of Edward Dewey and his Foundation for the Study of Cycles? We shall itemize the points covered in this chapter in order to make explicit the full ramifications of their research:

1) The discovery of cycles of the same periodicity (or, to employ their own physics analogy, wavelength) in so many areas of cyclic activity with no apparent connection to each other, such as wholesale prices and sunspot activity, implies a deep connection between physics and economics; it implies a kind of "econophysics";
2) The regularity of such cycles implied not only their inevitability but also their quantifiability and predictability;
3) This in turn implied that governmental policies could only exacerbate or alleviate the overall trend of a cycle, but never reverse or arrest it, *so long as the cycle remained a closed system without revolutionary expansion of the system itself, or introduction of radical new "Bottle-expanding" technologies.* For this reason, the underlying implied "econophysics" necessitates that the global elite has only three strategies open to it in the eventuality of achieving their goal of a global corporate state:
    a) Regionalized imbalances and controlled wars and conflict in

perpetuity, and various forms and degrees of population reduction, so long as the system remains closed in its "bottle";

b) an opening of the system by expansion of its environment, which practically can only mean a large human presence in outer space;

c) an opening of the system by a radical change in technology, a necessity if there is to be a large human presence in deep space.[73]

The last two alternatives pose some risk should the elite choose to follow them, for they inevitably carry with them the risk that their own power base can be inevitably challenged from a superior position, or superior technology, or both. Similarly, the first strategy carries the risk that their ability to control and manage such conflict would spiral out of their control.

4) The type of hyper-dimensional physics that their own words imply indicates a deep connection between

a) the harmonics of multiple-wave modulation or combination,

b) time,

c) cause, and

d) effect.[74]

5) This implies the most profound implication of them all: *if that physics were sufficiently known and a technology could be devised to tap into it, that selfsame physics would imply a technology to tap into, modify, and control human behavior itself, inclusively of the ability to manipulate aggregate economic behavior.*[75] One could, with a sufficient basis in this physics, directly manipulate whole economies and populations.

6) Not surprisingly, then, it was discovered that a large corporation and bank was represented in the Foundation for the Study of Cycles' membership,[76] and individual bankers appear to have utilized some sort of knowledge of such cycles to time their entrance into various markets, or the expansion of their financial empires.[77]

---

73 For this point, see the discussion in my book *The Philosophers' Stone: Alchemy and the Secret Research for Exotic Matter*, pp. 313–329.

74 This implies yet again the profundity of the temporal research of Russian astrophysicist Dr. Nikolai Kozyrev. Q.v. my *The Philosophers' Stone,* section three.

75 For this point, see the extended discussion in my *The Cosmic War: Interplanetary Warfare, Modern Physics, and Ancient Texts*, pp. 234–273 .

76 Dewey and Dakin also mention Bell Laboratories' own independent study of business and economic cycles. Q.v. *Cycles: The Science of Prediction*, p. 102.

77 Dewey and Dakin cite the instance of Wrigley and Rockefeller timing the building of large skyscrapers in Chicago and New York City to the bottom of a cycle, just when the overall trend was scheduled to go up again. Q.v. p. 122.

With all these conclusions and implications in mind, including especially their own suggestive analogy to the combination of several waves into one, then a profound question occurs: given the truthfulness of the analogy for the sake of argument, then *what is the carrier wave* for all these combined waves of cycles? What *is* the underlying reality, or the deeper physics, that they are suggesting? And for that matter, why does a certain class of the super-wealthy seem to *know* something about this "econophysics" that the rest of the business and economic world do not? And beyond the fact that some of that class might not know all of it, then why are so many of them apparently involved in the study of it?

The answers to these questions, surprisingly, lie in the mists, sands, texts, pavilions and columns of ancient history and temples, and a well-known and very ancient "pseudo-science": astrology.

Three

GERMANY, RCA, AND J.P. MORGAN
CASES OF INTEREST AND SUPPRESSION

∵

"The secret at the core of alchemy is an ineffable experience of the real workings of our local cosmological neighborhood."
—Jay Weidner and Vincent Bridges[1]

Dewey's and Dakin's sound wave analogy for their cyclic waves of economic activity raises the profound question: is there a *physics* basis to economic cycles? And if so, what is it? Moreover, if their own sound wave analogy is itself more apt than Dewey and Dakin themselves suspected, then what *carrier* wave modulates all these cycles of waves? As was seen in the previous chapter, the analogy of sound waves and carrier waves was no doubt suggested to Dewey and Dakin by the fact that such waves could be plotted together, or "averaged."

A. Dr. Hartmut Müller and Global Scaling Theory

But they were not the only ones to notice that waves of cycles of activity might point to a deeper physics connection. German physicist Dr. Hartmut Müller had considered a similar phenomenon from yet a different point of view, and proposed a theory to account for it: Global Scaling Theory.

Müller begins by noting that the theory is based upon what, for physics, is its primary "sacrament": measurement. But the centrality of measurement

---

1 Jay Weidner and Vincent Bridges, *The Mysteries of the Great Cross of Hendaye: Alchemy and the End of Time* (Rochester, Vermont: Destiny Books, 2003), p. 41.

in the methodology of physics soon reduced the physicist to playing the role of a kind of "court of arbitration" because his role is now that of deciding "which of the models" adopted by physicists to explain reality "matches the measurements and gets applied."[2] This assessment highlights the cul-de-sac in which physics deadlocked in the twentieth century, for there is no single testable theory valid over the *scales* or *sizes of objects* with which physics must deal, for the laws of standard physics operable at the scale of the very large, laws dominated by Einstein's relativity theories since the early twentieth century, are not applicable to the scale of the very small at the level of atoms and subatomic particles, where the laws of quantum mechanics prevail.

In this situation, physics and physicists are reduced, according to Dr. Müller, to being "mere interpreter(s) of models and ideas that got completely out of touch with reality — and this to an ever greater extent."[3] In short, like many other physicists, Dr. Müller is increasingly dissatisfied with the counterintuitive nature of contemporary physical mechanics, whether it be the relativity-dominated theories of the physics of large-scale systems or the quantum mechanical theories of the very small, with their emphasis on statistical probabilistic approaches.[4]

But there was one area in which the concentration on "the sacrament of measurement" did lead to something of major — if unappreciated — significance, and in this one discerns a direct conceptual connection to the work of Dewey and Dakin:

> The need for measurements of the highest precision promoted the development of mathematical statistics which, in turn, made it possible to include precise morphological and sociological data as well as data from evolutionary biology. Ranging from elementary particles to galactic clusters, this scientific database extends at least 55 orders of magnitude.[5]

---

    2   Dr. Hartmut Müller, "An Introduction to Global Scaling Theory," *Nexus*, Vol. 11, No. 5, September-October 2004, p. 49. See also the discussion in my *The Giza Death Star Destroyed* (Adventures Unlimited Press, 2005), pp. 118–123.

    3   Ibid., p. 49.

    4   For a discussion of two other physicists increasingly disenchanted with the counterintuitive nature of contemporary standard physical mechanics, see my *The Giza Death Star Destroyed*, pp. 130–150, for a discussion of the views of physicist Dr. Paul LaViolette and his "sub-quantum kinetics theory," and my *The Philosophers' Stone: Alchemy and the Secret Research for Exotic Matter* (Feral House, 2009), pp. 151–200 for a discussion of the torsion research of Russian astrophysicist Dr. Nikolai Kozyrev, who was similarly disenchanted with the results of relativity theory for the modeling of the energy of stars, and with quantum mechanics for its inability to give formally explicit and *engineerable* definitions of cause and effect. Kozyrev's work, as outlined by myself and others, led directly to the kind of "precursor engineering" of *causes* rather than *effects* in the top secret research of the Soviet Union, research which has continued apace in the Russian Federation.

    5   Müller, "An Introduction to Global Scaling Theory," *Nexus*, Vol. 11, No. 5, p. 49.

In other words, a vast database of measurements of all types and kinds of objects now existed, extending over the entire range of objects with which physics had to deal, from subatomic particles to planets, stars, and finally to whole galaxies and clusters of galaxies. With such a vast database of measurements at their disposal, one would think that scientists would have begun to contemplate its significance almost as soon as it had been available, namely in the early decades of the twentieth century.

Yet, notes Müller, this extensive database "did not become the object of an integrated (holistic) scientific investigation until 1982. The treasure lying at their feet was not seen by members of the labour-divided, mega-industrial scientific community."[6] Given the disenchantment of well-known and famous Russian physicists like Dr. Nikolai Kozyrev with the standard models of relativistic and quantum mechanical theories, it is not surprising that Russian scientists were the first to make a significant breakthrough and contribution toward the rise of Global Scaling Theory, for the first to draw attention to this database and its potential significance was a Russian biologist named Cislenko.

Publishing a scientific paper in Moscow in 1980 entitled "Structure of Flora and Fauna with Regard to Body Size of Organisms," Cislenko proved that "segments of increased species representation *are repeated on the logarithmic line of body sizes in equal intervals* (approximately 0.5 units of the decadic logarithm)."[7] In other words, Cislenko had discovered something very similar to Dewey's and Dakin's cycles of economic activity: *regular, periodic "wavelike" forms in the relative grouping of organic life's body sizes around certain values or measures of size.* There was just one problem: this grouping or "clumping" around certain scales of size "was not explicable from a biological point of view." For example, what mechanism could account for organisms preferring body sizes of "8–12 centimetres, 33–55 centimetres or 1.5–2.4 meters" and so on?[8]

For Hartmut Müller, however, the biological clumping around certain values in a periodic logarithmic scale pointed clearly to a deeper underlying physics basis for the phenomenon, for there were similar scale-invariant (i.e., applicable across all scales or measures of size) logarithmic distributions evident in the database of physics. Thus, by 1982, Müller was able to prove

> That there exist statistically *identical frequency distributions* with logarithmic, *periodically recurrent maximums* for the masses of atoms and

---

6 Ibid.
7 Ibid., emphasis added.
8 Müller, "An Introduction to Global Scaling Theory," *Nexus*, p. 49.

atomic radii as well as the rest masses and life-spans of the elementary particles."[9]

Shades of Dewey and Dakin! But that was not all, for very similar patterns were subsequently found for the logarithmic line of the "sizes, orbits, masses and revolution periods of planets, moons, and asteroids."[10] The phenomenon was scale-invariant, and, when graphed, revealed clear periodic "cyclic" types of "waves."

In other words, Dr. Müller was chin to chin with the same type of phenomena as those which confronted Dewey and Dakin, only in this instance the phenomena were purely based on *measurement of physical objects — particles and planets — and not economic cycles.* Yet the wave forms resulting from them were almost the same. In fact, in a certain sense, they were identical simply because they *were* waves!

This now puts Dewey's and Dakin's "sound wave analogy" and the question it poses in a whole new light, for it would now appear that we are no longer dealing with merely a conceptual analogy, but an analogy with actual correspondence in the real physical world. So the attendant question of that analogy — i.e., what is the *carrier wave* for all these diverse cycles? — also emerges into a whole new and vitally important context, and it is one Dr. Müller answers very directly, although he was probably unaware that he was also responding to a question posed by two American economists as a result of their own accumulation of a vast database of measurements.

1. *Longitudinal Waves in the Physical Medium*

Dr. Müller is a mathematical physicist, and as such was quick to perceive the relationship, since all the discrete measurements had a single unifying and underlying structure. That structure was simply that of a recurrent logarithmic period, i.e., a wave, or better, a harmonic series. It is when one considers the *cause* of this structure that the questions and implications begin to become evident and to multiply. As Dr. Müller states the case, the cause of this structure appeared to be *"the existence of a standing pressure wave in the logarithmic space of the scales/measures"* used to measure the phenomena themselves.[11] To put this point differently, I cite directly my comments about this observation from my *The Giza Death Star Destroyed:*

---

9   Ibid., p. 50, emphasis added.
10 Ibid.
11 Müller, "An Introduction to Global Scaling Theory," *Nexus,* p. 50, emphasis added.

> ...(T)he phenomenon of measurement itself, both as a phenomenon of intelligent observation and as a physical function, is scale-invariant, because regardless of the selected unit of measurement, the result still and always possesses a logarithm, a longitudinal standing wave structure.[12]

Note carefully, however, what Dr. Müller has actually said, and what he has *not* said. He has said that the phenomenon exists in a kind of "conceptual space." He has not said that the phenomenon exists *as a structure in the physical medium itself*, although it is clear that he is only a small step away from actually stating that. But as we shall now discover, he himself is inclined to the view that the phenomenon is physical in nature, and derives from a quantized view of the medium itself as a structured set of longitudinal "pressure" waves in that medium.[13]

As I have stated elsewhere,

> The results of this supposition were astounding, for regardless of the natural system studied, there were areas of "attraction," where any number of very discrete natural phenomena clustered, and "repulsion," areas avoided by natural systems. *So pervasive was this phenomenon that the German **Institute für Raum-Energie-Foeschung** (Institute for Space Energy Research) "was also able to prove the same phenomenon in demographics," i.e., in areas favored or shunned by urban populations, and in the economics both of nations and private businesses.*[14]

Note what is really going on here:

1) Müller's standing longitudinal waves in logarithmic space are applicable not only to physical phenomena, but to the same areas of activity as studied by Dewey and Dakin; thus,

---

12 Farrell, *The Giza Death Star Destroyed*, p. 120.
13 And this observation of course, calls to mind once again the hyper-dimensional theory of Müller's fellow countryman Burkhardt Heim, as well as the views of English physicist E.T. Whittaker and American Tom Bearden. In Heim theory, the physical medium, or space-time itself, begins in a state of absolute entropy or equilibrium. Even in that state, however, it is quantized, with the metric of the system being expressed as a single unit value, or "metron." As information increases in the system, the metron correspondingly is divided, or decreases. Similarly, as information increases, this is expressed as a quantized "rotation moment" for a particular area of space. Thus, Heim's view of space possesses certain nodes or lattice-like structures, and is thus commensurate with a view of space-time as a lattice-like structure of longitudinal waves in the medium itself. In a nutshell, space-time appears to "clump" around certain nodal points in that longitudinal wave's logarithmic structure. Lattice defects would on that view perhaps then emerge in real space-time as clusterings of matter or as physical forces.
14 Farrell, *The Giza Death Star Destroyed*, p. 120, emphasis added.

2) *Müller's Global Scaling Theory conclusively indicates that there is indeed a deeper physics to economic activity, to the very cycles of boom and bust as catalogued by Dewey's and Dakin's Foundation for the Study of Cycles;* and finally,

3) Note that the German Institute for Space Energy Research is sponsoring research *not only into this deeper physics, but doing so precisely in relation to a study of the physics of economic and human activity and social organization.* It is important for the reader to understand at this juncture that the term "Space Energy" in the German has the same sort of technical meaning or implication as the terms "vacuum" or "zero-point" energy has in English. In short, a German institution recognizes, funds, and researches not only Zero-Point energy, but researches it in the full range of its implications, including social and demographic, presumably for the purpose of engineering all of them to the extent possible.

But let us return for a few more insights from Müller's theory, for there is yet another significant implication that his theory holds.

That implication is *gravity*, for Müller's standing pressure wave in the medium may indeed be an explanation for the phenomenon of gravity. Simply put, the fact that natural systems "cluster" around the nodal points of these standing waves would indeed be a component or mechanism for the explanation of gravity in Müller's opinion.[15] It is important to understand the basic difference of this explanation from that of Einstein's General Relativity in terms of the order of concepts. For the latter, gravity is simply the curving of space-time that results from the presence of a large mass in local space-time. But Müller's theory is a much less superficial, and much deeper, theory, for *both* the warping of space-time *and* the large mass of a star or planet are both themselves the *results* of a much deeper structure around which both "cluster," and that structure is the structure of longitudinal pressure waves of compression and rarefaction in the physical medium itself. In a sense, the warp or curving of space-time is more fundamental than in Einstein's theory, for in a certain simplistic way of stating it, the warp results not from the presence of a large mass, but rather the converse: the large mass is present because of the underlying warp in the medium and the clustering of such waves around certain nodal points, almost the exact opposite of Einstein's theory.

This already gives one an appreciation of the significance of Global Scaling Theory, for in a certain sense, Einstein's theory is *not engineerable*, for if one

---

15   Müller, "An Introduction to Global Scaling Theory," *Nexus*, p. 52.

wished to warp space-time within the constraints of General Relativity, a large mass such as a star or planet would have to be present to do so. But the converse is true of Global Scaling theory, for one would not *need* the presence of a large mass to warp space-time, since that warped structure itself is a longitudinal pressure wave in the physical medium. One could, by warping space-time, create a "virtual mass" or, conversely, an "antigravity hill."

For Dr. Müller, there are two immediate practical applications of this theory. In the first instance, the theory of standing longitudinal waves in the medium opens the possibility of using such waves and *modulating them* — for they are indeed the mysterious "carrier waves" suggested by Dewey's and Dakin's sonic wave analogy — with information for communications purposes. Such waves would be much faster than ordinary light waves, since objects separated from each other over great distances in normal space can be quite close in the logarithmic space in which these waves move. Secondly, such a modulated "gravity wave" could be demodulated at any location on the Earth, Mars, or even outside the solar system, "at the very same moment in time."[16] The reason is simple enough. Large masses *are natural resonators of such waves.*[17] *And again, the reason is clear, for they tend to cluster around the nodal points or places where such waves interfere, that is, where they intersect.* Thus, the building of expensive satellites of space-based communications systems would not only be unnecessary, but in a certain sense, since such satellites are of much less comparative mass than a planet, they would not be cost-effective, since they would be very inefficient as coupled oscillators or receivers of such waves, as compared to a planet with its large mass itself.

2. The Link: Geometries

If there is a relationship between such longitudinal waves in the medium and gravity, it follows that as relationships of planetary alignments — the day-to-day, month-to-month, and year-to-year geometries of alignments of celestial bodies — will evidence similar relationships to human activity. A simple analogy will serve to illustrate this point. Imagine someone standing on the bank of a still pond, holding a handful of small pebbles. The individual then throws the rocks up, and they land in the pond, creating ripples and eventually, as all these waves cross and interfere with each other, a pattern or grid or *template* of wave interference is created, specific to

---

16 Müller, "An Introduction to Global Scaling Theory," *Nexus,* p. 82. For the relationship of the engineerability of the theory, see my discussion in *The Giza Death Star Destroyed,* pp. 122–123. Such a technology could also be used as a new and almost inexhaustible energy source, a propulsion technology, and of course a weapon of stupendous and planet- or star-busting power.

17 For this point, see my *SS Brotherhood of the Bell* (Adventures Unlimited Press, 2006), pp. 219–220.

that alignment of rocks thrown into the pond. After the pond surface has become smooth again, the individual throws another handful of pebbles into the pond. Again, the ripples will cross and interfere, but the pattern will be different because the rocks will randomly impact the surface of the pond in a different pattern.

The "ripples" are similar to the effects of these longitudinal waves in the physical medium, with one very important difference. In the case of the solar system, for example, the planets orbit the sun at known periodicity; that is, each planet takes a certain amount of time to complete one orbit of the Sun. This fact, the fact that their orbits can be precisely calculated and their positions relative to each other thus predicted, introduces the idea that certain types of alignments can occur with semi-regularity, and the physical effects thus can be predicted *if one assembles the requisite database from lengthy observation. That is to say, the planetary alignments themselves are exactly analogous to our "rocks in the pond," for the planets constantly create an overlapping gridwork or template of such longitudinal waves.*

a. Planetary Alignments and Signal Propagation

Oddly enough, such observation has been done to a limited extent. In the early 1950s, the RCA company engaged one of its engineers, J.H. Nelson, to study why signal propagation strength and weakness seemed to vary according to periodic cycles. Nelson's answers, after some study, were published in *The Electrical Engineer,* and are as shocking to "normal" scientific sensibilities now as they were then. The abstracts of these two short articles say it all.

In the abstract to his article "Planetary Position Effect on Short-Wave Signal Quality," Nelson announces

> A new approach to an as yet unsolved problem is the observance of planetary effects on trans-atlantic [sic] short-wave signals. Correlation over seven years shows that certain planetary arrangements agree well with the behavior of short-wave signals.[18]

In an article written only a few months earlier for the AIEE Winter General Meeting in January of 1952, Nelson spelled it out even more explicitly in the abstract, which stated

> An examination of shortwave radio propagation conditions over the North Atlantic for a five-year period, and the relative position of

---

18   J.H. Nelson, "Planetary Position Effect on Short-Wave Signal Quality," *Electrical Engineering,* May 1952, p. 421.

planets in the solar system, discloses some very interesting correlations. As a result of such correlations, *certain planetary relationships are deduced to have specific effect on radio propagation through their influence upon the sun.* Further investigation is required to fully explore the effect of planet positions on radio propagation in order that the highly important field of *radio weather forecasting may be properly developed.*[19]

There are some important factors to observe carefully before we proceed.

First, note that "certain planetary relationships" have an effect on radio signal quality and propagation via some influence they have on the Sun. This highlights that another mechanism other than sheer mass is at work, for as is obvious, the Sun's mass far outweighs the combined mass of the various planets! In other words, a standard relativistic model *could not explain why signal quality varied with planetary positions*; some *other* mechanism was at work than just the gravity-mass relationship posited by standard physical models. That mechanism, according to Nelson, had something to do with *the geometries of planetary positions over time.*

Secondly, Nelson also obliquely hints that in addition to the Sun's known effects on terrestrial weather, that there may also be a relationship between planetary positions themselves and terrestrial weather. Were it not for the fact that Nelson was an engineer working for RCA, one might be tempted to conclude that his article was that of some obscure astrologer working anonymously for *The Farmers' Almanac* making weather predictions.

But there's more.

To see what it is, it is necessary to follow Nelson's exposition closely and carefully. The problem first became evident when RCA erected a telescopic observatory to study sunspot activity:

> At the Central Radio Office of RCA Communications, Inc., in lower Manhattan, an observatory housing a 6-inch refracting telescope is maintained for the observation of sunspots. The purpose of erecting this observatory in 1946 was to develop a method of forecasting radio storms from the study of sunspots. After about one year of experimenting, a forecasting system of short-wave conditions was inaugurated based upon the age, position, classification, and activity of sunspots. Satisfactory results were obtained, *but failure of this system*

---

19 J.H. Nelson, "Shortwave Radio Propagation Correlation With Planetary Positions," conference paper presented to the AIEE Subcommittee on Energy Sources, AIEE General Winter Meeting, January 1952, www.enterprisemission.com/images.hyper/ne11.gif, p. 1.

from time to time indicated that phenomena other than sunspots needed to be studied. [20]

In other words, some mechanism other than sunspots was involved. But what was it?

Examining various articles by those who had studied the sunspot phenomenon in relationship to their cycles, Nelson quickly discovered something quite odd, and it figured significantly into his own discovery:

> Cyclic variations in sunspot activity have been studied by many solar investigators in the past and attempts were made by some, notably Huntington, Clayton, and Sanford, *to connect these variations to planetary influences.* The books of these three investigators were studied and their results found sufficiently encouraging to warrant correlating similar planetary interrelationships with radio signal behavior. However, it was decided to investigate the effects of all the planets from Mercury to Saturn, instead of only the major planets as they had done. *The same heliocentric angular relationships of 0, 90, 180, and 270 degrees were used and dates when any two or more planets were separated by one of these angles were recorded.*
>
> *Investigation quickly showed there was positive correlation between these planetary angles and transatlantic shortwave signal variations. Radio signals showed a tendency to become degraded within a day or two of planetary configurations of the type being studied.* However, all configurations did not correspond to signal degradation. Certain configurations showed better correlation than others.[21]

In other words, sunspot cycles themselves were not, as often argued, the cause of signal degradation, but were themselves correlated to planetary alignments; *both sunspots and radio propagation effects appeared to be effects of certain planetary geometries in relationship to the sun.* And there was another thing as well: the effect on signal propagation appeared *after* the geometrical alignment, which always occurred when two or more planets were separated by angles of 0, 90, 180, or 270 degrees.

Nelson's articles chart some of these planetary relationships, and it is worth having a look at them, for a picture is indeed worth a thousand words or a hundred equations.

---

20   J.H. Nelson, "Planetary Position Effect on Short-Wave Signal Quality," *Electrical Engineering*, May 1952, p. 421, emphasis added.

21   Ibid., emphasis added.

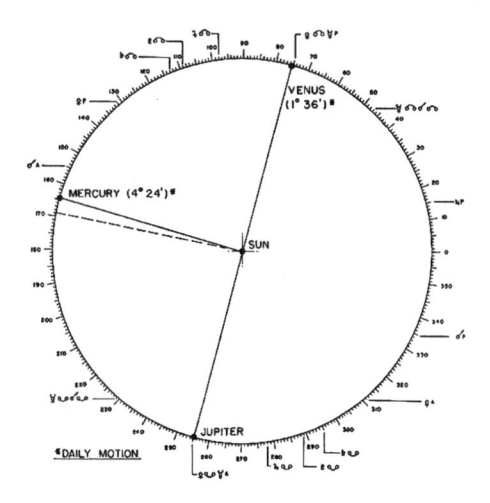

Nelson's Planetary Alignment Chart for February 23, 1948, which "resulted in severe signal degradation on that day and the one following."[22]

Note that Venus and Jupiter are separated by 180 degrees, and Mercury by almost 90 degrees from the other two. Yet another alignment that resulted in severe signal degradation was an alignment that occurred from September 20 to 26, 1951:

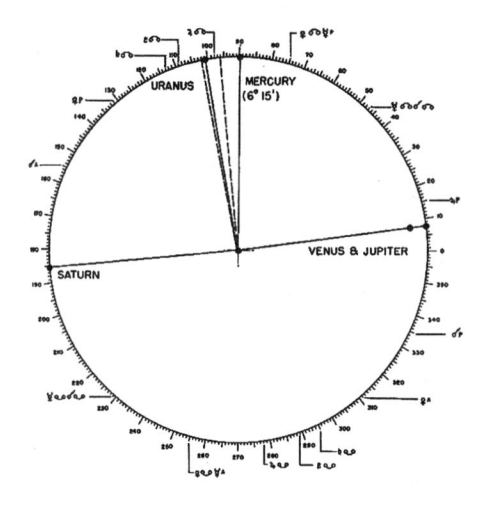

Nelson's Planetary Alignments for September 20–26, 1951[23]

---

22 J.H. Nelson, "Planetary Position Effect on Short-Wave Signal Quality," *Electrical Engineering*, May 1952, p. 422.

23 J.H. Nelson, "Planetary Position Effect on Short-Wave Signal Quality," *Electrical Engineering*, May 1952, p. 423.

Again, observe that Saturn on the one hand, and Venus and Jupiter on the other, are separated by almost 180 degrees, while Uranus and Mercury are within a few degrees of each other and almost at 90-degree angles of separation from the other three planets.

The similarity of these charts to normal astrological charts is, of course, obvious to anyone familiar with them, and it was so in Nelson's day as well, for *Time* magazine was quick to pick up on one potential implication of the study:

> The ancient pseudo-science of astrology, which attempts to predict the future by the motions of the planets, may have a bit of science in it, after all. This week Radio Corporation of America, no easy prey to superstition, announced in the RCA Review that it is successfully predicting radio reception by a study of planetary motions.[24]

To put it succinctly, *Time* was obliquely hinting at a kind of "paleophysics" approach to this ancient "pseudo-science," namely that behind ancient astrological myths and lore there might have once been a very sophisticated physics and science — the product of a Very High Civilization — of which astrology was a considerably declined legacy of civilizations, Egypt and Babylon for example, that were themselves declined legacies of a much earlier and much more sophisticated precursor civilization. And the astrological component, as we shall see momentarily, is more than just suggested or implied.

But we are getting ahead of ourselves.

For the moment, let us now consider carefully the implications of Nelson's remarks:

> 1) Since radio signals are electromagnetic phenomena, and since signal propagation appears to be affected by two or more planets in alignments of 0, 90, 180, or 270 degrees from each other, the implication is that the solar system is *not* electrically neutral at all, but an electrically dynamic and open system whose dynamism is affected by these alignments. Again, recall the idea that the planetary alignments themselves create a gridwork or lattice or template of longitudinal waves;[25]
>
> 2) The common feature of the Sun and the planets is that each is a rotating mass around its own axis of rotation, and as a total system,

---

24   No author given, "RCA Astrology," *Time,* Monday, April 16, 1951, www.time.com/magazine/article/0,9171,814720,00.html

25   For further discussion of the electrical dynamism of the solar system, see my *The Giza Death Star Destroyed,* pp. 29–31, and *The Cosmic War,* pp. 28–66.

each planet orbits the Sun. In short, we are dealing with rotating systems within rotating systems;

3) The Sun itself is a *rotating mass of plasma*, a super-hot electrically polarized gas in which nuclear fusion occurs constantly. Moreover, within the Sun's mass of rotating plasma, different layers of plasma north or south of the latitude of its equator *rotate in the same direction, but at different velocities*. This "differential rotation" as it is known will become a crucial consideration as we proceed.

But what does all this mean?

To answer this question, we must return to examine some aspects of this "deep physics" that I have written and commented about in my previous books. Many of these points can be reprised fairly quickly, but some of them will require a closer look, with new additional material and perspective.

b. The Electrically Dynamic Solar System and Planetary Alignments

As already noted, the RCA studies of J.H. Nelson point to the implication that the solar system is not an electrically neutral system, but a dynamic one whose changes are somehow connected to planetary alignments.[26] In most presentations of the electrically dynamic solar system, it is to be noted that the models used do *not* seek explanations in a "deeper" physics, but merely posit that in ancient times the planets were much closer together, and as a consequence of the buildup of polarities and charge differentials between planets, large-scale electrical arcs could form between them, much like charge differentials between the atmosphere and the ground form during large thunderstorms, leading eventually to the electrical arcs between these regions that we know as lightning. In this case, the electrically dynamic solar system's understanding of the way planetary alignments affect the system is simply due to the *relative closeness* of celestial bodies and the relative differences of charge polarities between them. The actual geometrical *alignment* of planets — as in the example of Nelson's RCA studies — is not really in view.

c. Plasma Cosmology

Closely allied to the conception of the electrically dynamic solar system is plasma cosmology, which is an attempt to see a deeper physics at work behind the idea of charge differentials forming between planets. With plasma cosmology and physics, we are indeed in the presence of a model with profound

---

26 Again, see my *The Giza Death Star Destroyed: The Ancient War for Future Science*, pp. 30–31; and *The Cosmic War: Interplanetary Warfare, Modern Physics, and Ancient Texts*, pp. 28–66.

implications for that "deeper physics," as well as a model with profound explanatory power for some ancient texts and mysteries.[27] Accordingly, our review must be more comprehensive.

I first outlined the basics, and the fundamental implications of plasma cosmology for interpretation of ancient texts and monoliths, in *The Giza Death Star*. The brainchild of Swedish physicist Hannes Alfvén, the conceptual basis of the model was summarized by his student Eric J. Lerner in the following fashion:

> Starting in 1936 Alfvén outlined, in a series of highly original papers, the fundamentals of what he would later term cosmic electrodynamics — the science of the plasma universe. Convinced that electrical forces are involved in the generation of cosmic rays, Alfvén pursued … (a) method of extending laboratory models to the heavens…. He knew how high-energy particles are created in the laboratory — the cyclotron, invented six years earlier, uses electrical fields to accelerate particles and magnetic fields to guide their paths. How, Alfvén asked, would a cosmic, natural cyclotron be possible?
>
> …But what about the conductor? Space was supposed to be a vacuum, thus incapable of carrying electrical currents. Here, Alfvén again boldly extrapolated from the lab. On earth even extremely rarified gases can carry a current if they have been ionized — that is, if the electrons have been stripped from the atoms…. Alfvén reasoned that such plasma should exist in space as well.[28]

I then commented at length as follows:

> This may not sound too revolutionary, until one notices what is unique about this theory: "certain key variables do *not* change with scale — electrical resistance, velocity, and energy all remained the same. Other quantities *do* change: for example, time is scaled as size, so if a process is a million times smaller, it occurs a million times faster."[29] In other words… *The principal differential is…time.* [30]

Note that idea that time is "the principal differential" or "thing in view," because it will become quite the crucial point in a moment.

---

27 Q.v. my *The Cosmic War: Interplanetary Warfare, Modern Physics, and Ancient Texts*, pp. 28–66.
28 Eric J. Lerner, *The Big Bang Never Happened* (New York: Vintage Books: 1992), p. 181, cited in my *The Giza Death Star*, p. 135.
29 Eric J. Lerner, *The Big Bang Never Happened*, p. 192, emphasis added, cited in my *The Giza Death Star*, pp. 135–136.
30 *The Giza Death Star*, pp. 135–136, emphasis added.

The principal insight of the theory is, however, clear: time is the primary differential or "thing in view" while the *rest* of physical variables — electrical resistance, field strength and so on — remain "scale-invariant," which is a physicist's technical terminology for saying that they behave the same no matter the size or scale of the system to which they are applied.

But there was a further implication:

> Since time is scale-sensitive, and other electromagnetic forces are not, the implication is revolutionary:
> "Equally important, though, is the converse use of these scaling rules. When the magnetic fields and currents of these objects are scaled down, they become incredibly intense — millions of gauss, millions of amperes, well beyond levels achievable in the laboratory. However, by studying cosmic phenomena, Alfvén shows, scientists can learn how *fusion devices more powerful than those now in existence will operate. In fact, they might learn how to design such devices from the lessons in the heavens.*"[31]
> ....
> Lerner clearly implies that *if the inertial and electromagnetic processes of the heavens are somehow captured...*

That is, if one is able to reproduce them by some technological means, then

> *"fusion devices more powerful than those in existence will operate."* What fusion devices could he be talking about? No tokamak magnetic bottle has ever achieved a stable controlled-fusion reaction, and it is unlikely that Lerner knows about (Philo) Farnsworth's plasmator...[32]

(And it is even less likely that Lerner knows of the Nazi Bell device!)[33]

---

31   Eric J. Lerner, op. cit., pp. 192–193, emphasis added, cited in my *The Giza Death Star,* p. 136.
32   For Philo Farnsworth's plasmator, see my *The Giza Death Star,* pp. 146–147, and *Nazi International,* pp. 328–333.
33   Considerations such as these have led some Nazi Bell researchers to the erroneous conclusion that the device was intended as an experiment in such mechanisms of controlled fusion for the purpose of creating a new energy supply as well as one of these "hydrogen bomb superbombs." As I noted in my *Nazi International,* the involvement of Dr. Ronald Richter in this project appears clear, and since his postwar project in Argentina ostensibly concerned itself with controlled fusion, such researchers reason back to the original project that the Nazi Bell was simply an earlier version of this postwar project. But Richter himself, as we shall see in a moment, indicates clearly that even this plasma research was a step or key to a much deeper and more fundamental physics. That such a plasma physics, however, might have led to such "fusion devices" would certainly have been a dividend of the project that the Nazis would have been only too happy to accept along the way! But it was merely a secondary dividend in any case.

...the only thing left, then, are the city-busting superbombs that fill French, American, and Russian arsenals.[34]

Those "city-busting superbombs" are, of course, thermonuclear hydrogen fusion bombs, and their importance to this story will also be seen in a moment.

But that's not all. There were other aspects of Alfvén's plasma cosmology model that he outlined in a 1942 article:

> If a conducting liquid is placed in a constant magnetic field, every motion of the liquid gives rise to an (electromagnetic field) which produces electric currents. Owing to the magnetic field, these currents give mechanical forces which change the state of motion of the liquid. *Thus a kind of combined electromagnetic-hydrodynamic wave is produced which, so far as I know, has as yet attracted no attention.* [35]

Commenting on this, I noted that such electromagnetic-hydrodynamic waves sounded suspiciously like the "electro-acoustic" or electrical longitudinal waves discovered by Nikola Tesla[36] in his high-frequency direct current impulse experiments, and later in his famous Colorado Springs electrical experiments. We shall have much more to say on this subject in a moment.

But there is one final point about Alfvén's plasma cosmology that must be noted, and that is that this concept of the universe

> Exhibits a filamentary and *cellular* structure. Not only does the universe exhibit "electric layers" of various densities like a fluid, but "cosmic plasmas are often nor homogeneous, but exhibit *filamentary structures* which are likely to be associated with currents parallel to the magnetic field.... In the magnetospheres there are thin, rather stable *current layers* which separate regions of different magnetization, density, temperature, etc. It is necessary that similar phenomena exist also in more distant regions. This is bound to give space a *cellular structure* (or more correctly, a cell wall structure)."[37]

Before continuing, it is essential that we pause and take stock of what we now have:

---

34 *The Giza Death Star,* pp. 136–137, emphasis in the original.
35 Hannes Alfvén, "Existence of Electromagnetic-Hydrodynamic Waves," *Nature,* No. 3805, October 3, 1942, pp. 405–406, cited in my *The Giza Death Star,* p. 137.
36 *The Giza Death Star,* p. 137.
37 Ibid., pp. 137–138, citing Hannes Alfvén, "On Hierarchical Cosmology," *Astrophysics and Space Science,* Vol. 89 (Boston: D. Reidel, 1983), 313–324, p. 314, all emphasis in the originals.

1) Alfvén's cosmology leads to the conception of electromagnetic-hydrodynamic waves which are capable of traveling in vacuum space, i.e., the physical medium itself;
2) The mechanism that attaches to these waves is plasma itself, which exhibits fluid-like properties of less and more dense regions, and thus,
3) Gives space itself a cellular, that is to say, a *lattice* structure;
4) The phenomenon of plasmas is scale-invariant, since plasma effects can be created in the laboratory which precisely resemble entire galaxies in their structure,[38] a point which may be fully appreciated by a glance at laboratory-generated plasma swirls, and their similarities to whole galaxies:

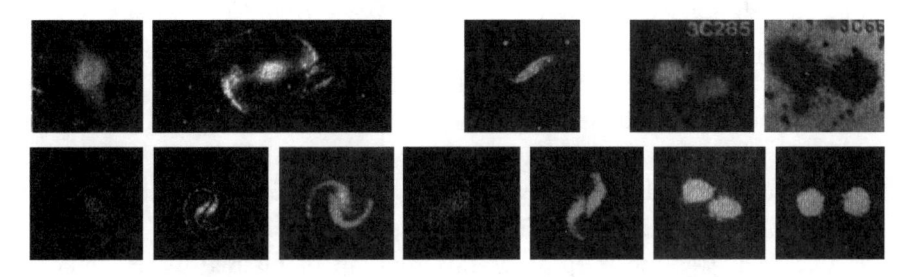

Galaxies and Laboratory Plasma Pinch Effects Compared: The Galaxies are above, and the Laboratory-Generated Plasma Pinches are below.

study of the above phenomena can lead to "fusion devices more powerful than those currently in existence," i.e., to super-super hydrogen bombs; and finally,

5) Time is the principal differential or "thing in view" since it alone varies over the scales which are in view; on small scales it operates faster than on large scales. This is an important point as we shall see in a moment in the work of Dr. Nikolai Kozyrev. However, given that time itself has this dynamic property considered *in relationship to the scale of the system under consideration*, it is clear that it cannot be the sort of "one-dimensional" or scalar entity that it is in the treatment of so much mathematical physics, particularly in relativity theories.[39] In short, time has a kind of "breadth and depth" similar to space; it is a *multi-dimensional* phenomenon, so to speak. And with this, we are

---

38 For this point, see my *The Cosmic War*, pp. 31–32.
39 For the "non-scalarity" of time in the work of Dr. Nikolai Kozyrev, see my *The Philosophers' Stone: Alchemy and the Secret Research for Exotic Matter* (Feral House, 2009), pp. 151–169.

back to the insights of Ouspensky cited by Dewey and Dakin in the previous chapter.[40]

With this in mind, we now consider the plasma and fusion aspects of the Nazi Bell device, and the postwar "plasma and fusion" research of Nazi scientist Dr. Ronald Richter, for his research was only secondarily about plasmas.

d. Plasma Transduction of the Vacuum or Zero-Point Energy: Dr. Ronald Richter Revisited

In *The Nazi International* I outlined a case that Dr. Ronald Richter's postwar fusion project in Argentina was but a continuation of certain aspects of the wartime Nazi Bell project. This case was based in part on the fact that Richter worked for the same company, the *Allgemeine Elektricitäts Gemeinschaft*, as built the power plant for the Bell device,[41] that while at that company he worked on a project code-named "Charite-Anlage *Projekt*," which was one of the code names of the Bell project's power plant,[42] but most importantly, that the physics involved in Richter's project and the Bell were so similar. Both the Bell and Richter's project involved the use of rotating plasma.[43] Both the Bell device and Richter's project also involved the use of metal drums or cylinders.[44]

But what was the purpose of this rotation of plasma? When queried about this precise point by an Argentine commission sent by President Perón to investigate his project, Richter informed the Argentine scientists that the whole basis of his attempt to control thermonuclear fusion processes was based on a precessional rotation of plasma that was then sharply and abruptly pulsed electrically to produce standing shockwaves within it. These

---

40 To draw a simple analogy, consider only human natural languages. That humans have always more or less known this "multi-dimensionality" of time is revealed by the way natural human languages — with their many tenses, moods, and active and passive voices — treat time. Verbal systems are, in short, far beyond the simple "past, present, and future" of conventional standard physical theories. Notably, human natural languages consider time in a highly complex set of *open systems and relationships between them*. Mathematical physics has yet to catch up with this subtlety in a certain sense.

41 Joseph P. Farrell, *Nazi International*, pp. 343–345.

42 Ibid., p. 345.

43 See my *SS Brotherhood of the Bell*, pp. 170–185, 278–282, 294–296, *Secrets of the Unified Field*, pp. 268–280, *The Philosophers' Stone*, pp. 283–287, 291–294, 296–305, and *Nazi International*, pp. 262, 314–315.

44 *SS Brotherhood of the Bell*, pp. 178–185; *Nazi International*, pp. 348–350. Harry Cooper of Sharkhunters, who photographed the drums of Richter's project in Argentina shown on p. 249 of my book *Nazi International*, has recently been informed by his Argentine contacts that the drums were in all likelihood control rod ports for Richter's intended fusion reactor that was beginning to be built when President Juan Perón shut down the project. This is possible, though it raises certain questions since *fission* reactors, not fusion reactors, are normally associated with control rods. This does, however, make some sense, as the drums in Argentina are solid weights with a hole through the center, while the Bell's counter-rotating drums were most likely hollow cylinders.

shockwaves would then induce the fusion reaction.[45] Richter was, of course, roundly denounced by scientists of that day, because such assertions flew in the face of the reigning theories.[46]

However, this does not tell the full story, for while Richter's views were being denounced publicly, privately and very secretly the U.S. military was taking a profound interest in what the Austrian Nazi scientist was *really* trying to do. Contacting Richter and asking him to explain what he was doing, Richter reveals that his project, and hence the entire Nazi Bell project, was only secondarily concerned with nuclear fusion, for that process was seen by them as a gateway to a much more profound, and potentially deadly, physics than mere fusion or even super-super hydrogen bombs. Outlining his research for the Third Reich, Richter informs his American investigators that the shockwave process of fusion was first noticed by him in *1936!* His research in Nazi Germany involved:

> Research work and design studies on electric arc furnaces systems, developing new types of plasma analyzing instruments and methods. In 1936, discovery of a plasma shock wave generating process, conception of a completely new type of industrial arc reactor system, based rather on plasma shock wave reactivity than on heat transfer. Development of a basis of operation for testing plasma shock wave conditions by means of plasma-collision-induced nuclear reactions.[47]

If all this sounds a little ahead of the thermonuclear game in the early 1950s, it was. It was about *45 years* ahead of that game, for in 1995 an American firm, General Fusion, announced it would try to build a fusion reactor based on the principle of plasma shockwaves![48]

But Richter had apparently seen something *else* in those early plasma shockwave and fusion experiments in Nazi Germany, something that opened the door to a much *deeper* physics. "We assume," Richter stated in his American Paperclip file,

> That highly compressed electron gas (i.e., a plasma) *becomes a detector for energy exchange with what we call zero-point energy...in a shock-wave-superimposed, turbulence-feedback-controlled plasma zone exists a high probability for cell-like super-pressure conditions...*

---

45   *The Nazi International,* pp. 262, 271–272.
46   Ibid., pp. 249–262.
47   National Archives and Records Administration, Foreign Scientist Case Files 1948-1958, Box 54 of Record Group 330, File on Dr. Ronald Richter, cited in Henry Stevens, *Hitler's Suppressed and Still-Secret Weapons, Science, and Technology* (Adventures Unlimited Press), pp. 260–261.
48   Farrell, *Nazi International,* p. 295.

Shades of Hannes Alfvén's plasma-cellular structure of space! But note what Richter is saying: such plasmas become "detectors" or even *gateways* to tap into the energy of the physical medium — the zero-point energy — itself:

> ...It seems to be possible (by this means) to 'extract' a compression-proportional amount of zero-point energy by means of a magnetic-field-controlled exchange fluctuation between the compressed electron gas and a sort of cell structure in space, representing what we call zero-point energy .... Plasma implosion analysis might turn out to become an approach to a completely new source of energy.[49]

In short, Richter has seen beyond Alfvén's plasma cosmology, for he is saying in effect that the plasma structure of space itself is a *result* of zero-point energy fluctuations. And with his "whole new energy source" based on plasma gateways into the zero-point energy, we are back to Eric J. Lerner's observation that by looking to the heavens and the plasma processes they contain, one might be able to construct fusion devices far more powerful than anything in today's arsenals, for such devices, in the final analysis, are not fusion devices as such but fusion "gateways" or transducers of something far more powerful and of planet- and star-busting potential.

So what is the connection to the Bell?

It's very simple.

Recall for a moment that Richter's whole process involved *rotation* of that plasma such that a precession or "wobble" was induced in it. Inside the Bell, there were two counter-rotating cylinders, in my opinion most likely stacked one on top of the other.[50] Inside these cylinders a highly radioactive compound of mercury, probably doped with high-spin-state nuclear isomers,[51] was spun up to ultra-high speed, on the order of tens if not hundreds of thousands of revolutions per minute. Isomers are in turn high-spin-state atoms whose energy is locked up in the angular momentum of the spinning atom. The purpose of spinning such a material *mechanically* up to such a high speed is therefore evident, for such action will "cohere" the atoms of the material by inertia, so that they are all spinning in a manner aligned more or less on the same *plane* of rotation, that of the two cylinders. This material was then *electrically pulsed* by extremely high voltage direct-current electricity[52] which, *pace* Richter, arc from the rotating drum itself, through the material, to the center of the device, thus

---

49 Farrell, *Nazi International*, p. 343.
50 Farrell, *Secrets of the Unified Field*, pp. 270–277.
51 Farrell, *SS Brotherhood of the Bell*, pp. 171–184, 294–295; see my *The Philosophers' Stone* pp. 297–299 for an argument that this isomer may have been thorium-229 isomer.
52 Farrell, *Secrets of the Unified Field*, pp. 280–282.

making not only a plasma, but setting up "electro-acoustic" shock waves within it, and these in turn, would transduce the energy and fabric of space-time itself, as the isomeric component of the compound deexcited from its high spin state and released the enormous amounts of energy stored up in its high rotation. By altering the speeds of counter-rotation of the two drums, a *differential rotation*, a precession or "wobble" of sorts, would be artificially induced in the plasma.

In short, the Nazi scientists had brought the plasma cosmology of the stars and even of the galaxies themselves down to a device as small as a small camping trailer, and were using it to manipulate directly the fabric of the physical medium, of space-time, itself.

But why set up such a differential rotation at all? For that matter, what factors might have led Dr. Richter to assume that a rotating (precessing) and electrically shocked plasma might be a zero-point energy transducer to begin with?

To answer that question, we must, once again, look to the stars, to the work of Russian astrophysicist Dr. Nikolai Kozyrev, and to little-known facts about the earliest hydrogen bomb detonations.

*e. Rotating Systems within Rotating Systems:
Dr. Nikolai Kozyrev Revisited*

To see what the connection of all this — from rotating plasmas to thermonuclear fusion to hydrogen bombs to stars and zero-point energy — really is, we must return to something I wrote way back in *The Giza Death Star*:

> During the first hydrogen bomb tests, the actual energy yield of the bombs far exceeded those initially calculated. There was an "x" factor, *an unknown source of surplus energy* that was being tapped. Since hydrogen bombs unleash enormous amounts of destructive energy at the very subatomic level of the nucleus of atoms, we may also surmise, in part, where that energy came from and why, for such weapons literally cause a violent local disturbance in the geometry and fabric of space-time. In short, *some as yet inadequately understood laws of harmonics produced the excess energy.*[53]

It is not only in the secret research of the Nazis with their Bell device, but in the work of Russian astrophysicist Dr. Nikolai Kozyrev that one begins the final approach to an understanding of where that excess energy came from and how it operates. We already know a part of the answer: it came from the energy of space-time itself.

---

53   Farrell, *The Giza Death Star*, p. 145, emphasis in the original.

But that space-time is *not* a mere void devoid of dynamism, for it contains *objects* and *systems*, it contains *information,* and those objects and that information are undergoing an ever-changing dynamic relationship to each other.

Kozyrev was confronted by a similar problem as the hydrogen bomb, only in his case, the problem was not with hydrogen bombs and their anomalous yields, but with the very Sun itself, and its own highly anomalous energy yield. As I wrote in *The Philosophers' Stone: Alchemy and the Secret Research for Exotic Matter:*

> The whole principal motivation, by his own reckoning, for his decades-long investigation of torsion, and of the torsion amplifying and shielding properties of various elements and compounds, was precisely the fact that there were simply not enough neutrinos being emitted by stars *for the standard model of stars as gigantic perpetual fusion reactors — essentially perpetual hydrogen bombs — to be true.* [54]

In other words, stars, like their man-made counterparts, the hydrogen bomb, were transducing *far too much* energy than could be accounted for by the process of thermonuclear fusion alone. In fact, in a certain sense, in Kozyrev's view, fusion was not even the primary energy source at work in stars at all, but a secondary *effect* of something much deeper, just as in the case of Richter's conceptions on plasmas. The connection between the two men's research, however, may not be immediately clear unless one recalls *what* stars really *are:* they are rotating balls of *plasma*. Thus, Richter's and Kozyrev's work are pointing to something significant: rotating plasmas are transducers of the energy of a higher-dimensional space! This is a crucial conceptual link to ancient times and beliefs, and a crucial key to why the international money power, even in ancient times, allied itself so closely with the temple, as we shall see in subsequent chapters.

And the similarity between Kozyrev and Richter is more than just conceptual, for both men and their theories were publicly denounced in the media organs of their respective political blocs, while their projects apparently disappeared into the labyrinth of postwar black projects:

> The more than coincidental attack on Kozyrev in *Pravda* in 1959 effectively [prevented] him from open publication of his experimen-

---

54 Farrell, *The Philosophers' Stone: Alchemy and the Secret Research for Exotic Matter* (Feral House, 2009), p. 193, referencing A.P. Levitch, "A Substantial Interpretation of N.A. Kozyrev's Conception of Time," p. 1. It should be noted that the "perpetual hydrogen bomb and neutrino emission theory" is yet another implication or artifact of Einstein's General Relativity theory, and, in that respect, is similar to all other such artifacts such as the Big Bang Theory, so-called "dark matter" and "dark energy" and so on.

tal results and theoretical conceptualizations, and the similarly-timed disappearance of discussion of clean fusion bombs in the open Soviet literature [is likewise suspicious]. Let us speculate a bit.

We have already encountered the fact that the earliest atmospheric hydrogen bomb tests far exceeded their pre-test calculated yields. In other words, *just as [Kozyrev maintained was happening] in stars, some other energy source was being tapped into, and transduced by, the thermonuclear detonation itself.* And if we extend this line of speculation, the Russians most likely encountered the same phenomenon in their hydrogen bomb testing. Moreover, in Kozyrev, they had an astrophysicist who thought he knew *why stars,* also implicating thermonuclear processes, appeared *not* to be radiating enough neutrinos energy for the thermonuclear model to be true.

One may reasonably and logically conclude, therefore, that the 1959 *Pravda* attack on Kozyrev was really a cover story to denounce his work, to *de-legitimize* it to anyone in the West who may have been paying attention to it, while Kozyrev, and his work, disappeared — as they did — into the highest reaches of classification within the Soviet Union, for his work provided the necessary key to understand why H-bombs were returning such anomalous yields, yields that, moreover, **most likely varied with the time of their detonation**. Kozyrev knew why: it was because the bomb itself became, for that brief brilliant nanosecond of the initial explosion, a dimensional gateway, a sluice-gate, opening the spillway to a hyper-dimensional cascade of torsion into the reaction itself.[55]

And when one says torsion, one says "rotation," that is, the geometry of time itself, with its rotating systems within rotating systems within rotating systems, for as Kozyrev throughout his many experiments demonstrated time and time again, the simplest physical phenomena — from hydrogen bomb fusion reactions to the simple inertial properties of gyroscopes and tension balances and scales — *varied in their results over time, i.e., in accordance with the position of celestial bodies in relationship to the earth, and to the position on the earth and the season in which an experiment was performed.* Kozyrev had in effect demonstrated that time — the geometrical configuration and alignment of systems in space at any given moment — was itself a physical force and factor that gave it breadth and depth and dimensionality. He had given the effects of so many cycles as compiled by Dewey and Dakin, and even the sunspot alignments of RCA's J.H. Nelson, a name: torsion.

---

55 Farrell, *The Philosophers' Stone,* pp. 194–195, italicized emphasis in the original, boldface emphasis added.

In stressing this geometric aspect of time and its "multi-dimensionality" — its relationship to the ever-changing torsion dynamism of objects in space, their "template" — Kozyrev could not possibly have been blind to what he was actually saying, for he was saying also that there may well be, or *have been*, long ago, a scientific basis to astrology, a basis now all but forgotten.

And there is one final point in this torsion tale to consider. The Sun, while a rotating plasma, is not a plasma that rotates *with uniform velocity*, for if one considers various layers of this plasma from its equator north, or south, as many solar physicists know, the plasma rotates at different speeds, i.e., the Sun itself is the largest example in our celestial neighborhood of the phenomenon of *differential rotation of a plasma*. As such, it is the example par excellence of torsion.

A simple analogy will illustrate this point. In numerous discussions on radio shows I have been asked to illustrate what torsion does to the fabric of space-time, of the physical medium. The example I always use is that of emptying an aluminum soda can, and then wringing it like a dishrag. The analogy is similar to the Bell's two counter-rotating cylinders, for the counter-rotation of the wringing motion will spiral, fold, and pleat the can, which represents space-time. But imagine now that one is wringing the soda can with both hands going the *same* direction, but with one hand going very much faster than the other: the result will be exactly the same. The can will be spiraled, folded, and pleated.

Thus, the analogy serves to demonstrate an important point both about the Nazi Bell research and about Kozyrev's research, for in *both* cases the torsion physics in evidence in plasmas was *fully* rationalized by both parties. Moreover, in stressing this geometric aspect of time and its "multi-dimensionality" — its relationship to the ever-changing dynamism of objects in space — Kozyrev could not possibly have been blind to what he was actually saying, for he was saying that there may well be, or *have been*, a scientific basis to astrology now all but forgotten.

There was, however, one genius in particular, working at the turn of the nineteenth into the twentieth century, and at the dreadful moment that the deliberate and heavy hand of suppression of this physics and all its implications could clearly be seen and felt, who saw it all with a perspicacity and breathtaking vision, and who inspired the efforts of the Nazis and Soviets to free themselves from that heavy hand:

Nikola Tesla.

B. All Roads Lead to Tesla and Morgan

One need only consider the conspicuous *absence* of Tesla's name from standard physics textbooks in order to appreciate the fact of the existence of this hidden physics and its deliberate suppression. It was evident in the public denunciations of Dr. Ronald Richter and Dr. Nikolai Kozyrev, and the quiet shuffling of their projects into classified obscurity and secrecy. But unlike those two geniuses, with Nikola Tesla one was dealing with an entirely different matter, for the man was not only *not* an obscure scientist and engineer working quietly in his laboratories on secret government projects, but he was a flamboyant and well-known public figure who delighted in large and expensive dinners in New York City's finest restaurants, who literally electrified audiences of the rich and famous — including such notables such as Mark Twain himself — with exhilarating public demonstrations of his inventions, and whose fame, quite literally, spread quickly worldwide with electrifying speed, for he indeed was the sole man responsible for electrifying the world after an equally famous contest with Thomas Edison and his financial backer J.P. Morgan — a contest which Tesla easily and decisively won.

Such a man could not simply be denounced, nor shuffled to the sidelines. Nor could such a man be dealt with by more "active measures" without casting a pall of suspicion over the ultimate perpetrators of the deed.

But why would Tesla pose such a problem to begin with?

1. Colorado Springs

The answer lies in what Tesla discovered during his famous experiments in Colorado Springs in the late nineteenth century, experiments which, to this day, few physicists and engineers really understand, and those few who do question either Tesla's claims, his analysis, or both. The reason for the difficulty in understanding among some scientists, and the questioning attitude and skepticism among others, lies in the nature of what Tesla himself claimed he had been able to do. It is necessary to cite his comments at length:

> Towards the close of 1898 a systematic research, carried on for a number of years *with the object of perfecting a method of transmission of electrical energy through the natural medium*, led me to recognize three important necessities: First, to develop a transmitter of great power; *second, to perfect means for individualizing and isolating the energy transmitted; and third, to ascertain the laws of propagation of currents through the earth* and the atmosphere.[56]

---

56 Nikola Tesla, "Transmission of Electrical Energy Without Wires," *Electrical World and Engineer*, March 5, 1904, cited in David Hatcher Childress, ed., *The Fantastic Inventions of Nikola Tesla*, pp. 219, 221, emphasis added.

The Colorado Springs project was, in other words, the beginning of Tesla's now well-known idea of beaming electrical power itself without wires. This is already a step well beyond radio.

Observe carefully, however, what Tesla states. In order to do this, it was necessary to

1) construct a kind of transmitter of extraordinary power;
2) to render any wireless transmission of power practical, it was necessary to be able to send and receive a multitude of individual signals for the various equipment and localities it was presumably to serve, and thus, a means had to be found to "individualize and isolate" the transmitted energy, much as a radio receiver can tune to various frequencies to receive different signals; and finally, and most importantly,
3) Tesla clearly states that the "natural medium" itself is to be the means of propagation of this energy.

But what does he mean by "natural medium"? The interpretation of this one point is crucial to the understanding of what he was really seeking in Colorado Springs, and it is this precise point over which the misunderstandings, arguments, and the skeptical questions, arise.

There are three possible ways to understand Tesla here. The first is to understand by "natural medium" what Tesla later declares in the same context: the atmosphere, and the earth itself. In the first instance, the atmosphere, Tesla is proposing little more than a high-power version of radio. In the second case, however, it is clear that he is proposing something radically different, for the earth itself is to be the transmitter. This interpretation, as we shall discover, is that which clearly will emerge from the rest of Tesla's remarks.

However, there is one final interpretation, the most radical of them all, that will loom ever larger as we proceed. Tesla, like many physicists and engineers of his day, was an ardent believer in the aether, that is, in the fact that space-time itself was a kind of ultra-fine matter upon which electromagnetic waves, and hence energy and power, could ride. Tesla, however, appears, unlike most physicists and engineers of his time, to have held the view that this aether had fluid-like properties, i.e., that it had the ability to be compressed or rarefied, that is, to be *stressed*. This will emerge as a clear implication of his Colorado Springs experiments.

To see how, we return to his own remarks, to understand how he intended to utilize the earth itself as the transmitting antenna for electrical power:

In the middle of June, while preparations for other work were going on, I arranged one of my receiving transformers with the view of determining in a novel manner, experimentally, *the electric potential of the globe and studying the periodic and casual fluctuations.* This formed part of a plan carefully mapped out in advance. A highly sensitive, self-restorative device, controlling a recording instrument, was included in the secondary circuit, while the primary was connected to the ground and an elevated terminal of adjustable capacity. The variations of potential gave rise to electric surgings in the primary; these generated secondary currents, which in turn affected the sensitive device and recorder in proportion to their intensity. *The earth was found to be, literally, alive with electrical vibrations,* and soon I was deeply absorbed in this investigation.[57]

On conclusion of these investigations, having discovered that indeed the earth itself was an electrically dynamic system, Tesla concluded that

Not only was it practicable to send telegraphic messages to any distance without wires, as I recognized long ago, but also to impress upon the entire globe the faint modulations of the human voice,

(In other words, Tesla *was* thinking in terms of radio, but with the entire earth as his transmitting antenna!)

[but] far more still, *to transmit power, in unlimited amounts, to any terrestrial distance and almost without any loss.*[58]

Tesla then turned to the problem of developing not only a powerful transmitter, but one that could best be used for all these discrete and disparate purposes. And it is precisely here that the misunderstandings begin, for most engineers to this day think that Tesla's "magnifying impulse transmitter" was nothing but a Tesla coil beaming powerful bolts of electricity into the atmosphere.

To see how different was Tesla's system than any other then being considered, one must look closer at what he himself describes happened in Colorado Springs. Noting the typical Colorado thunderstorms which formed over the mountains and then moved quickly over the plains, Tesla begins a lengthy description of the events leading to his discovery of the method of electrically stressing the earth itself:

---

57 Nikola Tesla, "Transmission of Electrical Energy Without Wires," *Electrical World and Engineer,* March 5, 1904, cited in David Hatcher Childress, ed., *The Fantastic Inventions of Nikola Tesla,* p. 222, emphasis added.
58 Ibid., pp. 226–227, emphasis added.

I never saw fire balls,⁵⁹ but as a compensation for my disappointment *I succeeded later in determining the mode of their formation and producing them artificially.*

In the latter part of the same month (June) I noticed several times that my instruments were affected stronger by discharges taking place at great distances than by those near by. This puzzled me very much. What was the cause? A number of observations proved that it could not be due to the differences in the intensity of the individual discharges,⁶⁰ and I readily ascertained that the phenomenon was not the result of a varying relation between the periods of my receiving circuits and those of terrestrial disturbances....

It was on the third of July — the date I shall never forget — when I obtained the first decisive experimental evidence of a truth of overwhelming importance for the advancement of humanity. A dense mass of strongly charged clouds gathered in the west and towards the evening a violent storm broke loose which, after spending much of its fury in the mountains, was driven away with great velocity over the plains. Heavy and long persisting arcs formed almost in regular time intervals....I was able to handle my instruments quickly and I was prepared. *The recording apparatus being properly adjusted, its indications became fainter and fainter with the increasing distance of the storm, until they ceased altogether. I was watching in eager expectation. Surely enough, in a little while the indications again began, grew stronger and stronger and, after passing through a maximum, gradually decreased and ceased once more. Many times, in regularly recurring intervals, the same actions were repeated until the storm which, as evident from simple computations, was moving with nearly constant speed, had retreated to a distance of about three hundred kilometers. Nor did these strange actions stop then, but continued to manifest themselves with undiminished force.... No doubt whatever remained: I was observing stationary waves.*

....Impossible as it seemed, this planet, despite its vast extent, behaved like a conductor of limited dimensions.⁶¹

In other words, Tesla, in Dewey- and Dakin-like fashion, had observed a *cycle*, and a very interesting one at that, for the cycle he had observed was

---

59   I.e., Ball lightning.
60   I.e., lightning.
61   Nikola Tesla, "Transmission of Electrical Energy Without Wires," *Electrical World and Engineer*, March 5, 1904, cited in David Hatcher Childress, ed., *The Fantastic Inventions of Nikola Tesla*, pp. 224–225, emphasis added.

a "stationary wave," or, in modern parlance, a "standing wave." When the extremely high-voltage static electrical discharges that we call lightning strike the earth, if the resonance is right it literally "thumps" the earth like a kettle drum, sending an impulse all the way around the planet somewhat like a seismic wave from an earthquake. Like all such waves, it will eventually fade and die, then grow again until it returns to the same place and intensity from which it started. And this gave him the idea for the "wireless transmission" of electrical power.

However, nothing could have been farther from Tesla's mind than standard radio. This new circuit is, he noted,

> Essentially, a circuit of very high self-induction and small resistance which in its arrangement, mode of excitation and action, *may be said to be the diametrical opposite of a transmitting circuit typical of telegraphy by Hertzian or electromagnetic radiations.... The electromagnetic radiations being reduced to an insignificant quantity*, and proper conditions of resonance maintained, the circuit acts like an immense pendulum, storing indefinitely the energy of the primary exciting *impulses and impressions upon the earth and its conducting atmosphere uniform harmonic oscillations* which, as actual tests have shown, may be pushed so far as to surpass those attained in the natural displays of static electricity.[62]

To unpack what he is saying, one only has to consider the standard radio transmission arrangement, sending out ordinary radio waves. In this arrangement, there is an antenna, connected to the ground. The antenna sends out Hertzian waves *into the atmosphere,* and the circuit is completed by its connection to the ground. Hertz waves are similar in nature to suspending a jump rope loosely between two people and having one person jerk the rope, sending an "S"-shaped wave slowly to the other end. Most of the energy of the wave is dissipated in the initial "jerking" motion, and accordingly, only a fraction of the energy arrives at the other person, the rest of the energy being dissipated by the up-and-down motion of the wave itself. In this case, the jump rope represents the earth's electrically conductive atmosphere, and the wave in the rope represents the ordinary radio wave.

Recall now what Tesla stated: his circuit parameters were diametrically *the opposite* of a standard "radio" arrangement, for in this instance, the relationship of ground and transmitter were turned on their heads, the earth, normally the

---

[62] Nikola Tesla, "Transmission of Electrical Energy Without Wires," *Electrical World and Engineer,* March 5, 1904, cited in David Hatcher Childress, ed., *The Fantastic Inventions of Nikola Tesla,* p. 227, emphasis added.

ground in an electrical circuit, became the *transmitter and conductive medium,* and the atmosphere, normally the conductive medium, became the *ground.*[63] Moreover, in this new arrangement, the waves were not Hertzian at all, but *impulses;* that is to say, they were longitudinal waves of stress, of compression and rarefaction, in the earth itself.

To understand the crucial difference between the two waves, we repeat our jump rope analogy, only this time, we place a yardstick between the two persons. One person pushes or pulses the yardstick repeatedly. Instantaneously, *all* of the energy of the pulse is transmitted directly to the other person. In this case, the yardstick represents the earth, and the pulses the electro-acoustic longitudinal waves Tesla is sending.

This is, to put it succinctly, a far cry from radio in the standard sense, for Tesla means not only to transmit signals by this means, but electrical power itself, without wires, for the earth itself *is* the wire.[64] Note also what the earth is in this system: it is a highly *non*-linear medium of almost infinite electrical "resistance" to a standard electrical current. This will assume some importance as we proceed.

In any case, once having envisioned this radically different system, Tesla went on to describe what for all practical purposes amounted to an "internet":

> The results obtained by me have made my scheme of intelligence transmission, for which the name of "World Telegraphy" has been suggested, easily realizable....Thus the entire earth will be converted into a huge brain, as it were, capable of response in every one of its parts.[65]

In fact, this is an "internet" beyond anything currently imaginable, for it has no need of wires, "gateway computers," or satellites in order to function. In Tesla's grand vision, the *earth* fulfills these functions.

2. Wardenclyffe

When he returned to New York City, Tesla was ready to put the scheme into commercial testing. And this is where Tesla's difficulties began, and where

---

63 For a fuller consideration of this point, see my *Giza Death Star Deployed,* pp. 197–205.

64 Tesla also states that "Simultaneously with these endeavors, the means of individualization and isolation were gradually improved. Great importance was attached to this, for it was found that simple tuning was not sufficient to meet the vigorous practical requirements." Once again, in other words, standard radio tuning techniques could not be utilized, since the waves were longitudinal "yardstick" waves of *pulses,* rather than Hertzian "jump rope" waves.

65 Nikola Tesla, "Transmission of Electrical Energy Without Wires," *Electrical World and Engineer,* March 5, 1904, cited in David Hatcher Childress, ed., *The Fantastic Inventions of Nikola Tesla,* pp. 230, 232, emphasis added.

suppression of the physics and engineering his system represented was clearly evident, for needing funds to build a large-scale version of his Colorado Springs experimental apparatus, Tesla, as is now well known, approached the American international banker J. Pierpont Morgan for a loan. Detailing his plan for a global system of "wireless telegraphy," even proposing to send the human voice and actual pictures by its means in a kind of terrestrial radio and television, Morgan advanced Tesla the funds. What Tesla did *not* reveal to Morgan at this juncture was that he intended to transmit electrical power *itself* by this means, making even his own stupendously successful system of alternating current — the system now in use worldwide — obsolete! Tesla chose a place on Long Island, purchased land, built a power plant, and a large tower. He called the place and the project "Wardenclyffe."

Unfortunately for Tesla and his plans, however, the Italian physicist and engineer Guigliomo Marconi — using Tesla's own devices and inventions to boot! — beat him to the punch by transmitting the first transatlantic radio signal, using standard radio circuit arrangements and Hertzian "jump rope" electromagnetic waves while Tesla's project was still under construction. Faced with cheaper competition and a deceptively simpler, but much less flexible, system, Morgan threatened to pull the financial plug on the scheme. It was then, faced with the demise of his project, that Tesla revealed his true intentions to Morgan: he planned to beam power itself through the earth to any point on the globe. With this revelation, Morgan *did* pull the plug, and Tesla's grand project came to an end, and has never since been publicly revived.

a. Weaponizing Wardenclyffe and Suppressing the Scalars:
Tesla at Tunguska and Morgan at Mischief

Why would Morgan pull the plug on such a scheme, virtually suppressing it and the physics it represented? After all, it is simply not true merely that he "could not meter" it and hence make money, for he stood to gain a great deal financially off of *worldwide* royalty and licensing agreements for the system, even though such a system would, admittedly, have been much more difficult to financially monitor and secure. So while pure and simple greed may very likely have been a factor in his deliberations, it probably was not the only one.

What those other modifications may have been were hinted at by Tesla himself, and a curious series of incidents that began when a French ship, the *Iena*, mysteriously exploded in 1907, as Tesla's Wardenclyffe project was in its financial death throes. In an op-ed piece written that March to the editor of the *New York Times,* Tesla hinted at the weaponization potential of his Colorado Springs discoveries and Wardenclyffe project:

> As to projecting wave-energy to any particular region of the globe, I have given a clear description of the means in technical publications. Not only can this be done by the means of my devices, but the spot at which the desired effect is to be produced can be calculated very closely, *assuming the accepted terrestrial measurements to be correct.* This, of course, is not the case. Up to this day we do not know a diameter of the globe within one thousand feet. My wireless plant will enable me to determine it within fifty feet or less, when it will be possible to rectify many geodetical data and make such calculations as those referred to with greater accuracy.[66]

A year later in 1908, the inventor was less guarded in his words:

> When I spoke of future warfare I meant that it should be conducted by direct application of electrical waves without the use of aerial engines or other implements of destruction. This means, as I pointed out, would be ideal, for not only would the energy of war require no effort for the maintenance of its potentiality, but it would be productive in times of peace. This is not a dream. Even now wireless power plants could be constructed by which any region of the globe might be rendered uninhabitable without subjecting the population of other parts to serious danger or inconvenience.[67]

These words were published in the *New York Times* on April 28, 1908. Bear that date in mind, because it will become highly significant in a moment. In any case, it is also clear that Tesla's wireless transmission of power was *one and the same technology as a horrible weapon of mass destruction.* And this raises an ominous series of questions: Had J.P. Morgan pulled his financial backing of the project not out of greed, but, perhaps motivated by some secret scientific advisor's cautionary warnings, had he pulled his financial backing because he did not wish to see technology with such destructive potential in the private hands of a scientist well known for very odd and eccentric behavior? Were Morgan's motivations ultimately altruistic? Or, conversely, did he want to develop the technology for such purposes secretly himself, and by means of its potential use, establish a world mastery for himself and his own class?

To put it succinctly, as with the presence of the Mitsubishi bank economist in the membership roles of the Foundation for the Study of Cycles, or the

---

[66] "Tesla's Wireless Torpedo: Inventor Says He Did Show that it Worked Perfectly," *New York Times,* March 19, 1907.

[67] "Mr. Tesla's Invention: How the Electrician's Lamp of Aladdin May Construct New Worlds," *New York Times,* April 21, 1908.

peculiar investigations of RCA engineer J.H. Nelson, this is the *third* clear instance of a major corporation, secure in its own monetary temple, showing a clear interest, first in the development of a deep physics, a radically different physics, and now, the first clear indication of a deliberate act of financial sabotage of the same physics! This pattern will only become more and more acute as we proceed throughout this book.

But in that very same article, Tesla makes clear that he has seen beyond his own system of using the earth itself as an antenna to an *even deeper* underlying physics, a physics he did *not* disclose at the time to Morgan but which now, desperate for financial support for his project, he now disclosed publicly in the pages of the *New York Times*:

> What I said in regard to the greatest achievement of the man of science whose mind is bent upon the mastery of the physical universe, was nothing more than what I stated in one of my unpublished addresses, from which I quote: "According to an adopted theory, every ponderable atom is differentiated from a tenuous fluid, filling all space *merely by spinning motion, as a whirl of water in a calm lake. By being set in movement this fluid, the ether, becomes gross matter. Its movement arrested, the primary substance reverts to its normal state. It appears, then, possible for man through harnessed energy of the medium and suitable agencies for starting and stopping ether whirls to cause matter to form and disappear.* At his command, almost without effort on his part, old worlds would vanish and new ones would spring into being. *He could alter the size of this planet, control its seasons, adjust its distance from the sun, guide it on its eternal journey along any path he might choose through the depths of the universe. He could make planets collide and produce his suns and stars, his heat and light; he could originate life in all its infinite forms. To cause at will the birth and death of matter would be man's grandest deed, which would give him the mastery of physical creation, make him fulfill his ultimate destiny.* [68]

Note now that by "medium" Tesla no longer means simply the earth itself, but the aether. And notably, he views atoms themselves as but vorticular structures — rotating, torsion-based systems — *in* that aether. If Morgan had any doubts in 1907 about the potentials — both destructive and constructive — of Tesla's system and the physics it embodied, those doubts were clearly dispelled a year later by Tesla himself in *The New York Times*.

---

68 "Mr. Tesla's Invention: How the Electrician's Lamp of Aladdin May Construct New Worlds," *New York Times*, April 21, 1908, emphasis added. Tesla's letter to the *Times* was dated April 19, 1908.

Note also that Tesla, by viewing the physical medium in this torsion-based fashion, has also clearly seen one implication: the aether itself was a transmutative, *alchemical* information-creating medium, a medium which, as information-creating, was in a state of non-equilibrium and broken symmetry. Thus, with the proper structuring of that "torsion potential," by *engineering it*, one could literally alter the size and mass of the earth or any other celestial body at will, for those bodies were themselves "whirls in the aether." Small wonder then that he, like Drs. Richter and Kozyrev after him, was creating ball lightning and thereby studying plasmas and electrical shocks, for like them, he has seen the much deeper physics of the physical medium itself, and like them, he has discovered the basic method of manipulating and engineering it through electrical pulses or longitudinal waves, and through rotation. And like them and their Nazi and Communist masters, he has seen the full range of its constructive and destructive implications. In this, and without the slightest exaggeration, his views of physics gave a solid basis to a very ancient pseudo-science: alchemy.

But why is the timing of Tesla's announcement so significant?

One researcher who asked this question was Oliver Nichelson:

> The question remains of whether Tesla demonstrated the weapons application of his power transmission system. Circumstantial evidence found in the chronology of Tesla's work and financial fortunes between 1900 and 1908 points to there having been a test of this weapon.[69]

Recall that Tesla's first veiled hints of the weaponization potential of his wireless power transmission system were stated in 1907 and clearly expressed by April 1908. Roughly two months later, on June 30, 1908, a massive explosion occurred over Tunguska, Russia, in Siberia.

> An explosion estimated to be equivalent to 10–15 megatons of TNT[70] flattened 500,000 acres of pine forest near the Stony Tunguska River in central Siberia. Whole herds of reindeer were destroyed. The explosion was heard over a radius of 620 miles. When an expedition was made to the area in 1927 to find evidence of the meteorite presumed to have caused the blast, no impact crater was found. When the ground was drilled for pieces of nickel, iron, or

---

69 Oliver Nichelson, "Tesla Wireless Power Transmitter and the Tunguska Explosion of 1908," prometheus.al.ru/english/phisik/onichelson/tunguska.htm, p. 6. Note that the website posting Mr. Nichelson's paper is Russian!

70 America's first hydrogen bomb, codenamed "Mike," was estimated to have a yield of 4–6 megatons prior to detonation. When detonated, the explosion ran away to about 10 megatons, approximately the same yield as the *minima* estimation for the Tunguska event.

stone, the main constituents of meteorites, none were found down to a depth of 118 feet.⁷¹

Many explanations have been given for the Tunguska event. The officially accepted version is that a 100,000-ton fragment of Encke's Comet, composed mainly of dust and ice, entered the atmosphere at 62,000 m.p.h., heated up, and exploded over the earth's surface creating a fireball and shock wave but no crater. Alternative versions of the disaster see a renegade mini-black hole or an alien space ship crashing into the earth⁷² with the resulting release of energy.⁷³

As was already seen, Tesla had two short months earlier spelled out the weaponization potential of his magnifying transmitter device and its underlying deep physics very clearly.

Many years later in 1934, according to a letter unearthed by Tesla biographer Margaret Cheney, the inventor once again wrote J.P. Morgan:

> The flying machine has completely demoralized the world, so much so that in some cities, as London and Paris, people are in mortal fear from aerial bombing. The new means I have perfected affords absolute protection against this and other forms of attack… *These new discoveries I have carried out experimentally on a limited scale, created a profound impression.* ⁷⁴

At this point one might ask, what prompted the letter? Was it spontaneous on Tesla's part? Or a reply to an inquiry from the financier? Whatever lay behind this strange correspondence, occurring years after the financier had severed his connection with Tesla and all but suppressed the Wardenclyffe wireless power transmission project, the correspondence is odd to say the least.

Its implications however, are clear. Tesla's remarks in the *New York Times* in 1908, the same year as the event, and his letter years later in 1934 to Morgan, point to a circumstantial case that he indeed tested the weaponization potential of his wireless power transmission on a remote wilderness region, a region where there would be sufficient observation of the event from a distance,

---

71   All this raises the question of why the Soviet Union would *mount* such an expensive expedition in the first place to such a remote wilderness area. One possibility, of course, was they were looking for clues of whether or not the event was a natural one, or artificially induced and therefore a product of a technology.

72   Always the inept ET incompetently crashing his ships into the earth! All that technology, and such *bad* pilots! Don't they have flight simulators and pilot schools? After all, *we* do.

73   Nichelson, op cit., here summarized in different form by David Hatcher Childress, *The Fantastic Inventions of Nikola Tesla*, p. 256.

74   Childress, *The Fantastic Inventions of Nikola Tesla*, p. 256, emphasis added.

but no loss of human life. Indeed, in purely legal terms, he had "the motive and the means to cause the Tunguska event. His transmitter could generate energy levels and frequencies capable of releasing the destructive force of 10 megatons, or more, of TNT. And the overlooked genius was desperate."[75]

But as Nichelson also observes, the nature of the Tunguska explosion is also entirely consistent with what would happen

> with the sudden release of wireless power. No fiery object was reported in the skies at that time by professional or amateur astronomers as would be expected when a 200,000,000-pound object enters the atmosphere at tens of thousands of miles an hour. Also, the first reporters, from the town of Tomsk, to reach the area judged the stories about a body falling from the sky was the result of the imagination of an impressionable people. He noted there was considerable noise coming from the explosion, but no stones fell. The absence of an impact crater can be explained by there having been no material body to impact. An explosion caused by broadcast power would not leave a crater.
>
> In contrast to the ice comet collision theory, reports of upper-atmosphere and magnetic disturbances coming from other parts of the world at the time of and just after the Tunguska event point to massive changes in earth's electrical condition. Baxter and Atkins cite in their study of the explosion, *The Fire Came By,* that the *Times* of London editorialized about "slight, but plainly marked, disturbances of…magnets"….
>
> In Berlin, the *New York Times* of July 3rd reported unusual colors in the evening skies thought to be Northern Lights…Massive glowing "silvery clouds" covered Siberia and northern Europe. A scientist in Holland told of an "undulating mass" moving across the northwest horizon. It seemed to him not to be a cloud, but the "sky itself seemed to undulate." A woman north of London wrote the London *Times* that on midnight of July 1st the sky glowed so brightly it was possible to read large print inside her house. A meteorological observer in England recounted on the nights of June 30th and July 1st: "A strong orange-yellow light became visible in the north and northeast… causing an undue prolongation of twilight lasting to daybreak on July 1st…. There was a complete absence of scintillation or flickering, and no tendency for the formation of streamers, or a luminous arch, characteristic of auroral phenomena….Twilight on both of these nights was prolonged to daybreak, and there was no real darkness."

---

75 Oliver Nichelson, "Tesla Wireless Power Transmitter and the Tunguska Explosion of 1908," prometheus.al.ru/english/phisik/onichelson/tunguska.htm, p. 9.

...

When Tesla used his high-power transmitter as a directed-energy weapon he drastically altered the normal electrical condition of the earth. By making the electrical charge of the planet vibrate in tune with his transmitter he was able to build up electric fields that affected compasses and caused the upper atmosphere to behave like the gas-filled lamps in his laboratory. He had turned the entire globe into a simple electrical component that he could control.[76]

But was a "simple" wireless transmission of power *all* that was involved in the explosion? After all, we have already found Drs. Richter and Kozyrev questioning the standard models of thermonuclear fusion, and Tesla himself by 1908, clearly seeing a deeper physics of the medium itself, was involved in his own wireless power transmission principles.

More importantly, flattening an area of 500,000 acres is equivalent to flattening an area of 781.25 square *miles,* or roughly an area of a square bounded by sides almost 28 miles long! So some other mechanism might be in play, and Tesla has already told us what it might have been: whirls in the aether itself — *torsion, rotation.*

However, Tesla had foreseen even further political implications for his system, and it is worth citing him at length:

> Through the universal adoption of this system, ideal conditions for the maintenance of law and order will be realized, *for then the energy necessary to the enforcement of right and justice will be normally productive, yet (potentially) and in any moment available, for attack and defense. The power transmitted need not be necessarily destructive, for if existence is made to depend upon it, its withdrawal or supply will bring about the same results as those now accomplished by force of arms.*
>
> *But when unavoidable, the same agent may be used to destroy property and life.* The art is already so far developed that great destructive effects can be produced at any point on the globe, determined beforehand and with great accuracy. In view of this I have not thought it hazardous to predict a few years ago that the wars of the future will not be waged with explosives but with electrical means.[77]

---

76 Oliver Nichelson, "Tesla Wireless Power Transmitter and the Tunguska Explosion of 1908," prometheus.al.ru/english/phisik/onichelson/tunguska.htm, pp. 9–10. it should be noted that Nichelson thinks that Tesla originally was aiming for the North Pole arctic region with this demonstration, and by a miscalculation, overshot his target and hit Siberia instead.

77 Nikola Tesla, "Tesla's New Device Like Bolts of Thor: He Seeks to Patent a Wireless Engine for Destroying Navies by Pulling a Lever; to Shatter Armies Also," *New York Times,* Dec. 8, 1915, p. 8, column 3, emphasis added.

In other words, like the "gods" of extremely ancient times,[78] Tesla saw in his technology of wireless power transmission a means of world mastery and hegemony, whose implied threat of withdrawal of access, or even worse, the implied threat of its destructive weaponized deployment, would coerce compliance to whatever world order as possessed, monopolized, and deployed it. Such statements could hardly have gone unnoticed by Tesla's former financial backer, J.P. Morgan.

But whatever Morgan's interest may have been, foreign nations were most certainly also watching these statements carefully. As Oliver Nichelson notes, the American media carried many stories of "death ray" research being conducted *in the wake of Tesla's statements to the New York Times* in Russia,[79] France,[80] and Great Britain.[81]

In the last instance, Tesla biographer John J. O'Neill noted in an unpublished chapter to his celebrated biography of the inventor, *Prodigal Genius*, that when he queried the engineer about his wireless power transmission technology and its weaponization potential, Tesla became very reluctant to speak about it in any detail. But later, Tesla did offer some interesting information:

> Somewhat later I learned the reason for Tesla's reluctance to discuss the details. This came shortly after Stanley Baldwin replaced Neville Chamberlain as Prime Minister of Great Britain [sic].
>
> Tesla revealed that he had carried on negotiations with Prime Minister Chamberlain for the sale of his ray system to Great Britain for $30,000,000 on the basis of his presentation that the device would provide complete protection for the British Isles against any enemy approaching by sea or air, and would provide an offensive weapon to which there was no defense. He was convinced, he declared, of the sincerity of Mr. Chamberlain and his intent to adopt the device as it would have prevented the outbreak of the then threatening war, and would have made possible the continuation — under the duress which this weapon would have made possible — of the working agreement involving France, Germany and Britain to maintain the status quo in Europe. When Chamberlain failed, at the Munich conference, to retain this state of European equilibrium it was necessary to get rid of Chamberlain and install a new Prime Minister who could make

---

78 For the whole theme of an ancient cosmic war and the use and deployment of such technologies of "cosmic hegemony," see my *Giza Death Star Deployed*, pp. 1–4, and my *The Cosmic War: Interplanetary Warfare, Modern Physics, and Ancient Texts*.

79 Oliver Nichelson, "Tesla's Wireless Power Transmission and the Tunguska Explosion of 1908," citing *The New York Times*, "Suggests Russia Has A 'Ray,'" May 28, 1924, p. 25.

80 Ibid., citing *Current Opinion*, "A Violet Ray That Kills," June 1924, pp. 828–829.

81 Ibid., citing *Popular Mechanics*, "'Death Ray' Is Carried by Shafts of Light," August 1924, pp. 189–192.

the effort to shift one corner of the triangle from Germany to Russia. Baldwin found no virtue in Tesla's plan and peremptorily ended negotiations.[82]

However, there is a large looming problem over hovering over these assertions, and that is simply the sequence of historical facts.

Neville Chamberlain remained Prime Minister until the Nazi invasion of the Low Countries and France in May of 1940, and did *not,* as is alleged, step aside for Stanley Baldwin to return to Britain's premiership after the Munich Conference of 1938. In fact, Chamberlain remained, even after Churchill had replaced him, on Churchill's War Cabinet until his death a few months later from cancer. The entire sequence of events is wrong, for it was Chamberlain who had replaced *Baldwin* as prime minister in 1937. And in any case, it is unlikely that Chamberlain, who pursued a policy of appeasement with Nazi Germany, would have been interested in a weapon of mass destruction that, had the Germans caught wind of the plan as they inevitably would have, would have exacerbated tensions between the two nations even more.

But need one discard these allegations merely for the presence of this (admittedly whopping) historical mistake?

Perhaps not, for it is possible that O'Neill is simply recording accurately what *Tesla* told him, and Tesla, then already a very old man, might have not remembered things in their precise sequence, confusing *Chamberlain* with *Baldwin.* On this reading, admittedly only a possibility, Tesla's deal was being negotiated with Stanley Baldwin, and the deal was subsequently scrubbed by Chamberlain on his assumption of the premiership and in the interests of appeasement.

All this is mentioned because, of course, there is one country conspicuously *absent* from the previous roll call of countries involved with "death ray" research in the 1920s: Germany. However, as researcher Oliver Nichelson observes, the Germans were indeed working on such a weapon, for the *Chicago Tribune* reported the following interesting little story:

> Berlin – That the German Government has an invention of death rays that will bring down airplanes, halt tanks on the battlefields, ruin automobile motors, and spread a curtain of death like the gas clouds of the recent war was the information given to Reichstag members by Herr Wulle, chief of the militarists in that body. It is learned that three inventions have been perfected in Germany for the same purpose and have been patented.[83]

---

82 Cited in www.tfcbooks.com/articles/tunguska.htm, pp. 6–7.
83 Oliver Nichelson, "Tesla's Wireless Power Transmission and the Tunguska Explosion of 1908,"

Self-evidently, such stories abounded in the 1920s in the wake of Tesla's pronouncements, but there are sound reasons to take them seriously, especially in the case of Germany.

Under the terms of the Versailles Treaty, Germany was prohibited from development of heavy artillery over certain calibers, and absolutely prohibited from having tanks or even an air force, and its standing army was limited to a mere 100,000 men. However, the Treaty did not prohibit development of these types of weapons, and thus it is logical to assume that Germany in particular would seek to do a technological end-run around the restrictions of the treaty.[84]

Even more interesting, however, is the fact that Nichelson reports that one British researcher allegedly involved in the development of such systems, J.H. Hamil, observed that the German system was based "on an entirely different principle" than those in evidence in other reports.[85] While Hamil explicitly states that his own "death ray" is based on Tesla's Colorado Springs wireless power transmission experiments,[86] it is also interesting that it appears that he *misinterpreted* the nature of those experiments, for he built a large Tesla coil and was apparently attempting to beam power by its means through the atmosphere, the exact *opposite*, it will be recalled, of what Tesla was trying to do. This makes his statements all that much more interesting, for by pursuing "an entirely different principle" were the Germans in fact using the *earth* as the transmitter, and the atmosphere as the ground, as Tesla himself indicated? We will never know, but it is interesting to note that whatever they were doing, it was not the same as what everyone *else* was doing, according to Hamil.[87]

---

citing May 25, 1924 as the date of the *Tribune* article. The story was also apparently reported in *The New York Times*, "Suggests Russia Has A 'Ray,'" May 28, 1924, p. 25.

84   It is also worth noting that the year is significant — 1924 — for it was the same year that Nobel laureate physicist Prof. Dr. Walther Gerlach wrote openly in the *Frankfurter Allgemeinezeitung* that a large-scale project should be initiated for the investigation of "alchemical"-like transmutations the Germans were observing in scientific experiments. Q.v. my *SS Brotherhood of the Bell*, pp. 272–278. Germany also investigated more directly related technologies as is alleged here, and this research continued during the period of the Third Reich: q.v. my *Secrets of the Unified Field*, pp. 239–248.

85   Oliver Nichelson, "Tesla's Wireless Power Transmission and the Tunguska Explosion of 1908," p. 3.

86   Ibid.

87   Researcher Brian Desborough presents rather interesting, though unsubstantiated, information concerning the relationship between Morgan and Tesla: "Disturbed by Tesla's command of free-energy and anti-gravity technology, Tesla's financial backer, Louis Cass Payseur, ordered his front man J.P. Morgan to withdraw his financial support. This act financially ruined Tesla through the initiation of false foreclosure proceedings.

"....another added impetus for Morgan's financial withdrawal was the amorous advances made toward Tesla by Morgan's daughter. The manservant previously loaned Tesla by Morgan was, in reality, an industrial spy who, in collusion with one of Tesla's staff, a German engineer named Fritz Lowenstein, stole Tesla's anti-gravity secrets and assigned them to various Illuminati secret societies, both in Britain and America." (Brian Desborough, *They Cast No Shadows* [Writers Club Press, 2002], p. 295.)

b. Lt. Col. Tom Bearden on Scalar Resonance

But what *is* precisely the relationship between Tesla's use of the earth itself as his transmitter, and the far more dangerous idea he himself suggests, that such "standing waves" as he observed at Colorado Springs, could be used as a means of engineering and manipulating the physical medium itself for whatever purpose, constructive or destructive?

The answers to this question are provided in certain suggestive statements Tesla made in a trial transcript during a trial in New York state. Tesla, now in dire financial straits after J.P. Morgan stopped all further backing of the Wardenclyffe wireless power transmission project, had given the deed of his Wardenclyffe property to the owner of a hotel where he resided as security in return for him being allowed to room in the hotel. When Tesla was unable to pay the hotel bill the owner naturally enough foreclosed on the property, winning a judgment against Tesla. Tesla appealed the case, and it is this transcript that contains some interesting descriptions of the Wardenclyffe property by Tesla, and some even more interesting allusions to the physics properties he thought to be behind it. I cite the transcript at length:

By Mr. Hawkins:

> Q. Were there any other structures upon the premises other than that brick factory or laboratory which you have just described?
>
> A. (Tesla speaking) Yes sir, there was the structure which in a certain sense was the most important structure, because the power plant was only an accessory to it. That was the tower.
>
> Q. Please describe the tower as to dimensions and material and method of construction and kind of construction?
>
> > Mr. Fordham: We renew our objection, if the Court please. This is entirely immaterial, irrelevant and incompetent until after they have succeeded in establishing their contention that the deed is a mortgage.
> > The Referee: I will take it.
> > Mr. Fordham: Exception.
>
> A. The tower was 187 feet high from the base to the top. It was built of special timber and it was built in such a way that every stick could be taken out at any time and replaced if it was necessary. The design of the tower was a matter of considerable

difficulty. It was made in the shape of an octagon and pyramidal form for strength and was supporting what I have termed in my scientific articles a terminal.

By the Referee:
Q. There was a sort of globe at the top?
A. Yes. That, your Honor, was only the carrying out of a discovery I made that any amount of electricity within reason could be stored *provided you make it of a certain shape. Electricians even today do not appreciate that yet.* But that construction enabled me to produce with this small plant many times the effect that could be produced by an ordinary plant of a hundred times the size. And this globe, the framework was all specially shaped, that is the girders had to be bent in shape and it weighed about 55 tons.

By Mr. Hawkins:
Q. Of what was it constructed?
A. Of steel, all the girders being specially bent into shape.
Q. Was the tower that supported it entirely constructed of wood or partly of steel?
A. That part alone on top was of steel. The tower was all timbers and of course the timbers were held together by specially-shaped steel plates.

The Referee: Braces?
The Witness: Yes, steel plates. I had to construct it this way for technical reasons.
The Referee: We are not interested in that.

Q. Was the tower enclosed or open?
A. The tower, at the time of the execution of this deed, was open, but I have photographs to show how it looked exactly and how it would have looked finished.
Q. After you delivered the deed was the tower ever enclosed?
A. No, it was just open.
Q. Now the dome or terminal at the top, was that enclosed?
A. No sir.
Q. Never enclosed?
A. Never enclosed, no.
Q. *Had that structure ever been completed?*

A. The structure so far, if I understand the terms right, yes, the structure was all completed but the accessories were not placed on it yet. For instance that globe there was to be covered with specially pressed plates. These plates —

Q. That had not been done, had it?

A. That had not been done, although I had it all prepared. I had prepared everything, I had designed and prepared everything, but it was not done.[88]

Q. Was the structure of the tower in any manner connected with the brick building or power plant?

A. The tower was separate.

Q. I understand, but was there any connection between them?

A. There were of course two channels. One was for communicating, for bringing into the tower compressed air and water and such things as I might have needed for operations, and the other one was to bring in the electric mains.

By the Referee:

Q. In order to do that there was, as a matter of fact, was there not, a well-like shaft going down right in the middle of the tower into the ground some 50 or 60 feet?

A. Yes. You see the underground work is one of the most expensive parts of the tower. In this system that I have invented it is necessary *for the machine to get a grip of the earth, otherwise it cannot shake the earth. It has to have a grip on the earth so that the whole of this globe can quiver,* and to do that it is necessary to carry out a very expensive construction. But I want to say this underground work belongs to the tower.

By Mr. Hawkins:

Q. Anything that was there, tell us about.

A. There was, as your Honor states, a big shaft about ten by twelve feet, goes about 120 feet and this was first covered with timber and the inside with steel and in the center of this there was a winding stairs going down and in the center of the stairs

---

88 This is an important point and argues *against* Nichelson that Tesla used his Wardenclyffe project to make a demonstration of its capabilities at Tunguska. Thus, Tesla would have had to use another facility. However, note that Tesla also indicates that the basic preparations for the completion of the tower had been completed. We are left then with a mystery and a question: could the facility have been duplicated somewhere and by someone else?

there was a big shaft again through which the current was to pass, *and this shaft was so figured in order to tell exactly where the nodal point is, so that I could calculate every point of distance. For instance I could calculate exactly the size of the earth or the diameter of the earth and measure it exactly within four feet with that machine.*

Q. And that was a necessary appurtenance to your tower?

A. *Absolutely necessary.* And then the real expensive work was to connect that central part with the earth, and there I had special machines rigged up which would push the iron pipes, one length after another, and I pushed these iron pipes, I think 16 of them, 300 feet, and *then the current through these pipes takes hold of the earth.* Now that was a very expensive part of the work, but it does not show on the tower, but it belongs to the tower.

....

Q. Tell the Court generally, not in detail, the purpose of that tower and the equipment which you have described in connection with it?

> Mr. Fordham: How is that material?
> The Referee: I will take it.
> Mr. Fordham: We except.

A. Well, the primary purpose of the tower, your Honor, was to telephone, to send the human voice and likeness around the globe.

By the Referee:

Q. Through the instrumentality of the earth.

A. Through the instrumentality of the earth. That was my discovery that I announced in 1893…

....

By Mr. Hawkins:

Q. The purpose then briefly was for wireless communications to various parts of the world?

A. Yes *and the tower was so designed that I could apply to it any amount of power and I was planning to give a demonstration in the transmission of power which I have so perfected that power can be transmitted clear across the globe with a loss of not more than five per cent, and that plant was to serve as a practical demonstration....*[89]

---

89 New York State Supreme Court, Appellate Division, Second Department: Clover Boldt Miles and George C. Boldt, Jr. as Executors of the Last Will and Testament of George C. Boldt, Deceased, Plaintiffs-Respondents, versus Nikola Tesla, Thomas G, Shearman, et al. as Defendants-Appellants, 521–

Now look what we have:

1) At the same approximate time as Tesla is making these statements in a trial, he is making statements for the *New York Times* that indicate the same technology of wireless power transmission is capable of using its longitudinal waves to manipulate the physical medium — not just the earth — itself, either for constructive or massively destructive purposes;
2) In the transcript itself, Tesla is very careful to state that not only is his system for communications purposes, but for the express purpose of the transmission of *power in any amount*;
3) To do this, Tesla also clearly states in the transcripts that the Tower structure at Wardenclyffe also had a deep shaft so that his system could physically and electrically "grip the earth" and "grip the entire globe" and make it "quiver"; and to do this,
4) Tesla also clearly implied that it was necessary to have as exact a measurement of the earth's size, of its *geometry*, as possible.[90] In other words, the geometrical and non-linear properties of the earth itself were essential components to the system.

The reason it was essential may be recalled from our previous remarks with respect to Dr. Harmut Müller's Global Scaling Theory, namely, that longitudinal pressure waves in the medium itself are responsible for the local warping of space-time and the clustering of objects near certain nodal points where such waves overlap. To put it differently, large masses such as planets or stars *are the natural and most efficient resonators of these waves*:

> (This) scalar coupling of the solar system provides a major check on unrestricted use of ... large...strategic scalar EM weapon systems. If significant scalar effects are produced on earth in a "pulse" mode, *pulsed disturbance of the earth-sun and earth-moon system results*. Here a danger exists that one or more natural resonances of the coupled systems may be excited. If the feedback stimulation of the Sun is not insignificant, for example, large sunspot activity may result sometime thereafter, say in a day or two. If too much or too sharp stimulation occurs on earth, the coupled resonant response from the sun could be disastrous....

537, pp. 174–179, cited in David Hatcher Childress, ed., *The Fantastic Inventions of Nikola Tesla* (Adventures Unlimited Press), pp. 314–319, all emphasis added.

90 For those really paying attention, q.v. my *The Giza Death Star*, pp. 239–248, esp. p. 248, and *The Giza Death Star Deployed*, pp. 149–169.

Accordingly, use of huge scalar (electromagnetic) weapons is a double-edged sword. Unless carefully employed, use of the weapons could cause a terrible backlash to the user as well as the victim, and even accidentally cause the destruction of the earth itself.[91]

In other words, *precise* knowledge of local celestial geometries, including the earth, is at all times essential in such a system as Tesla envisioned, and this is, perhaps, why he himself ultimately described the implications of his technology for a deeper physics able to engineer the physical medium itself destructively or constructively. And note, in Tesla's vision, it was quite essential that the system itself be physically and electrically rooted to the earth, for the earth indeed was a huge component in a gigantic electrical circuit. Thus, they do greatly err who would maintain that such a system would more practically be deployable on satellites in space, for the whole point of such longitudinal *waves of stress in the medium itself is that they are best established by manipulating their natural resonators: whole planets and stars.* To say otherwise is entirely to misunderstand the physics.[92] The *best* place for the deployment of such systems is precisely on the surface of planets.

C. Conclusions

So at the end of this journey, we have the following things:

1) There clearly exists a physics of open systems in which ever-changing celestial geometries exercise mutual physical influences;
2) This physics points to a deeper physics of the physical medium itself, which is manipulable via longitudinal waves of stress within the medium, which in turn may be accessible via methods of electrical stress in nonlinear mediums such as the earth, or, in the case of stars, rotating plasmas;
3) This physics and the ability to incorporate whole planets as components of a technology clearly existed, and was asserted, as early as the turn of the last century, in the work and words of Nikola Tesla, who

---

91 Tom Beaden, *Fer De Lance,* pp. 209–210, cited in my *SS Brotherhood of the Bell,* pp. 219–220.

92 Such, alas, is the case with the recent appearance of the work of a friend of mine, Sesh Heri, and his *Handprint of Atlas*. Heri indeed advances and argues a novel and intriguing hypothesis, namely, that the entire topography and geomorphology of land features and even the continents themselves are results of such standing waves in the earth. And in a brilliant insight, he couples this idea with the whole idea of the layout of certain artificial monuments and constructions around significant naturally occurring geomorphologies. These brilliant insights remain worthy of consideration, despite my following critique of his misunderstandings.

Heri states in his book the following in reference to my weapon hypothesis for the Great Pyramid as outlined in my *Giza Death Star Trilogy*:

noted both its benign and malign implications were both logical and simultaneous entailments of that same physics and system of technology;

4) As the database of physical science expands with increasing measurements over a vast and disparate body of knowledge, correlations between seemingly unrelated systems began to be noticed, and these correlations included cycles of human economic, sociological, and emotional activity, suggesting the possibility of a *deep correlation between physics, economics, and celestial geometries*;

5) Significant indicators were also found of discrete but definitive corporate and financial interest in the discovery of the principles of these connections between physics and economics, and in the case of Tesla and Morgan, of efforts to *privately develop* and later *suppress* its instrumentalities.

It is in those last two points — the deep correlation between physics, economics, and celestial geometries, and the private development and suppression of its instrumentalities — that we discover a host of questions, and a very ancient, indeed, "paleoancient" connection....

"Such a thesis as Farrell develops in his Giza Death Star trilogy is compelling, breathtaking, chilling, and scientifically convincing in its details, and yet....

"The same critique that I would apply to Dunn's electrical power plant thesis, I would also apply in a variant form to Farrell's weapon thesis as well. While the Great Pyramid, as a transmitter of longitudinal electric waves, could certainly be deployed as a weapon, why would it have been built specifically for the purpose of being used as a weapon? Is a ground-based longitudinal electrical wave transmitter the ideal way to configure a massively destructive scalar beam weapon — especially a scalar beam weapon designed for interplanetary war? I would think that a mobile position in space would be the strategically ideal location for a scalar beam weapon.... A civilization that could build the Great Pyramid could also build a giant space platform and mount upon it a longitudinal electric wave transmitter that could propagate massively powerful scalar beams capable of splitting a planet apart. Such a space platform would also be capable at an instant's notice of moving from its position in space at superluminal speed....

"So why would an advanced civilization build a weapon that was a "sitting duck"? Much more likely, a facility such as the Great Pyramid would be an object to be protected by a space platform, not primarily a weapon. Certainly, the Great Pyramid would have been capable of propagating scalar beams in defense if it was attacked. But a true predatory weapon would have to be able to move — and move very quickly."

Before proceeding with the rest of Heri's critique of my weapon hypothesis for the Great Pyramid, these critiques must be addressed. As already noted in the main text, the best natural oscillators of these "scalar" resonances are large masses such as planets and stars. To miss this point is to miss entirely the point of the physics involved. Thus, if one *were* to build a mobile space platform for Heri's version of such a weapon, one might accordingly have to do so on an almost planetary scale. Not that it could not be done, of course, for the types of civilizations as he and I are positing might be capable of doing so, but the purpose of doing so is not cost-effective, for it is *entirely* unnecessary for such a weapon system to be able to *move* in order to *target* a specific object, no matter how distant. The targeting is by resonance and interferometry. The "point-aim-and-shoot" model he is implying is not necessary and indeed in a certain sense almost counter to the whole nature of the system. Similarly, such a system would be defensible precisely on the same basis of exploiting the base planetary system's geometry and local resonance. In short, it is not necessary for it to *move* in order to be either offensively "predatory" or "defensible" in either case.

Heri's second critique is far more telling, but suffers from its own shortcomings:

"But this strategic flaw is not the most compelling argument against Farrell's view that the Great Pyramid was originally destined to be a weapon; rather, it is Farrell's own repeated emphasis that the Great Pyramid is too "over-engineered" to be anything other than a weapon that draws my attention. Farrell makes the point several times in his Giza trilogy that the evidence indicates that the Great Pyramid was able to transmit energy to "Any Possible Receiver in local space" without requiring a complex receiving apparatus at the load end. And he presents some very good evidence and arguments that this indeed was the case. But from this he extrapolates that such a sophisticated device could only be constructed for one purpose and one purpose only — to be used as a weapon of mass destruction — to decimate cities and continents — or to blow up planets. But while such a sophisticated device certainly could be used as a weapon, there is no logical reason that it could only be used as a weapon. Indeed, the very kind of highly sophisticated and sensitive tuning characteristics of such a device suggest to me a more likely constructive use than a destructive one. Sensitive, precise focusing and tuning are the characteristics of careful, constructive manipulation of material substance, not a sledge-hammer disintegration of it. A device that could send energy to "Any Possible Receiver" could send much more than destructive waves — it would have the capacity to send carefully modulated waves of energy that could communicate, nourish, build, and grow. The extreme precision evinced by the Great Pyramid strongly suggests that it was designed to be used as a scalpel, not a sword."

While I am grateful to Heri that he has indeed perceived my arguments accurately and outlined them fairly to a certain extent, and has not resorted to the dismissive tactic of some in the alternative community who simply dismiss the whole hypothesis out of wishful thinking or appeals to "mentors" and "authorities" with access to a presumed millennia-old unwritten tradition, again Heri fails to see the nature of the physics involved, for precision tuning of the very finest surgical precision would indeed be needed — as evidenced by Tesla's own remarks — *especially* if it were deployed as a weapon, precisely in order to avoid the feedback resonance effects noted by Bearden. So this cannot be used as an argument against the weapon hypothesis.

Similarly, the reverse argument that Heri implies — that all its scalpel-like precision and fine-tuning evince a constructive purpose for its construction and primary use also fails, for as the copious citations of Tesla in the main text I hope make clear, *one and the very same* physics and technology can be used for *either* purpose, that is to say, anyone building any such device would *know* as an inevitable implication of the physics employed and involved that one was building both a "power plant" *and* a weapon. It is impossible to disentangle the two, and in Tesla's own words, in such a system it is necessary to "grip the earth." I have merely stated the case in the *Giza Death Star Trilogy* for the weapon hypothesis as being the ultimate purpose of the Pyramid's construction simply because so many wish or choose to ignore this implication of the physics involved. But it is clearly evident in the physics. The true purposes of building such a system would, needless to say, probably be kept secret from the general public at any time, just as Tesla kept the true purpose of Wardenclyffe hidden from J.P. Morgan, and even after revealing it to him, only *later* and in desperation publicly disclosed its weaponization potential. But that potential was clearly present from the beginning as a logical entailment of the physics, and anyone building such a system would know that. Therefore to exclude the possibility of an intentional construction of a weapon from the motivation or purpose in building such systems is *a logical impossibility*. In short, at one and the same time as one is building a system for all Heri's wonderful constructive purposes, one is building by the nature of the case, a weapon, whether one builds it in space or not, and at the very moment that system becomes operational for all those wonderful and constructive purposes, it also becomes operational as a weapon.

(Quotations from Sesh Heri, *The Handprint of Atlas: The Artificial Axis of the Earth and How it Shaped Human Destiny* [Highland, California: Corvos Books, Lost Continent Library Publishing Co., 2008] pp. 240–241.)

II.

The Temples, The Stars, and the Banksters

"...so it was that the temple that should owe fealty to the gods alone, became a front for the international money creative force of that day and age; connected closely with the trade in precious metals and slaves as it must have been."
—David Astle, *The Babylonian Woe*, p. 25.

Four

Temples, Templates, and Trusts
The Ancient Roots of a Deep Relationship

❖

> *"The determination of the stars to which some of the Egyptian temples, sacred to a known divinity, were directed, opened a way, as I anticipated, to a study of the astronomical basis of parts of the mythology."*
> —J. Norman Lockyear[1]

> *"...to put it in the terse language of Bastiat, society implies exchange, and exchange, money."*
> —Alexander Del Mar[2]

Paradoxically, the farther back in history one goes, the closer the relationship between science, magic, and money becomes. Many have commented upon this relationship, but few have understood its significance for the type of physics that it implies. The nature of this relationship, and the conspiratorial implications that it implies, is best exhibited by examining the relationship of gold mining and slavery in ancient Egypt. But it is important to begin at the beginning, in order to see how, and in what manner, the actions of a conspiracy can be glimpsed. From earliest times, the power to "make and issue money and regulate the value thereof" (to paraphrase the U.S. Constitution) has been recognized to be the sole prerogative of the State, or, in more ancient times, the crown or king.

---

1 J. Norman Lockyear, *The Dawn of Astronomy: A Study of Temple Worship and Mythology of the Ancient Egyptians* (Dover Publications, 2006), p. xvi.
2 Alexander Del Mar, *A History of Money in Ancient Countries from the Earliest Times to the Present* (Kessinger Publications, reprint of the George Bell and Sons edition, 1885), p. 15.

As the famous nineteenth-century American numismatic scholar Alexander Del Mar aptly put it, "The right to coin money has always been and still remains the surest mark and announcement of sovereignty."[3] Indeed, it is because it is a sure "mark and announcement of sovereignty" that one finds the association of the making and regulation of the legal value of money so intertwined not only with the sovereignty of the state or crown, but with the sovereignty of God (or, as the case may be, of the gods) from whom the power of rule in most ancient societies was invariably perceived to derive, and no better exemplar of this complex relationship between the state control of money issuance, of mining, and the close interlock of both with religion can be found than that of ancient Republican and Imperial Rome.

A. Temples and Trusts
1. The Roman Model

As money in ancient times was primarily — though certainly not exclusively — thought to reside in precious metals such as gold, copper, silver, and in some cases even bronze, it inevitably followed that for a state to maintain its sovereignty over the making, issuance, and regulation of the value of money, it had to exercise a strict control over the mining of those metals:

> Therefore, the means necessary to secure and maintain such a money were for the State to monopolize the copper mines, restrict the commerce in copper, strike copper pieces of high artistic merit, in order to defeat counterfeiting, stamp them with the mark of the State, render them the sole legal tenders for the payment of domestic contracts, taxes, fines, and debts, limit their emission until their value (from universal demand for them and their comparative scarcity) rose to more than that of the metal of which they were composed, and maintain such restriction and over-valuation as the permanent policy of the State. For foreign trade or diplomacy a supply of gold and silver, coined and uncoined, could be kept in the treasury.
>
> There are ample evidences that means of this character were, in fact, employed by the Roman Republic; and, therefore, that such was the system of money it adopted. The copper mines were monopolized by the Roman State, the commerce in copper was regulated, the bronze nummi were issued by the State, which strictly monopolized their fabrication, and the designs were of great beauty, the pieces were stamped "S.C.," or *ex senates consulta*....their emission was limited,

---

[3] Alexander Del Mar, *History of Monetary Systems* (Honolulu: University Press of the Pacific, 2000), p. 66.

until the value of the pieces rose to about five times that of the metal they contained, and they steadily and for a lengthy period retained this high over-valuation.[4]

Thus, we have three interconnected ideas at the outset:

1) The power to coin and regulate the value of money by established law was a prerogative solely of the state or crown, and not of any private money- or credit-issuing monopoly;
2) There was no inherent value in "precious metals" as such, but such value had to be artificially created by the state issuing such money by two means:
    a) the relative scarcity of precious metal money issued, which made the legal monetary value of money issued exceed the inherent value of the metal in the coin itself; and,
    b) the value was further enhanced by the creation or addition of artwork to the coins issued; and,
3) Because the issuance of money was connected to precious metals such as copper, silver, and gold, the state or crown had to maintain a monopoly over the mining and stockpiling of such metals.

a. The Bullion Trust and the Temple

But in ancient Rome, there is a further factor at work, and it is best to allow Del Mar himself to state what it is: "It is impossible to resist the conviction," he states, "that the superior value of gold in the West *was created by means of* legal and, perhaps, also *sacerdotal ordinances.* This method of fixing the ratio may even have originated in the Orient."[5] In other words, in fixing the value of gold at a ratio of certain units of value — expressed in terms of other metals — per unit of gold, gold's value as the ultimate precious metal was established largely through religious ordinances. Moreover, Del Mar suggests this practice originated in "the Orient," which could mean, as far as Rome was concerned, the Middle East or even the East as far away as India. So one now must add, to the above list, a fourth connection:

4) The value of gold was fixed largely by "sacerdotal ordinance," in other words, *there is a deep and profound connection between the issuance of money, its fixation of value in law in terms of precious metals whose value in turn is defined by ratios per unit of gold on the one hand, and*

---

4 Alexander Del Mar, *A History of Monetary Systems,* pp. 21–22.
5 Ibid., p. 87, emphasis added.

religion on the other. In short, one is in the presence of the deep connection of ancient conceptions of money, and religion, or as I put it in the chapter title, between "temples" and "trusts."

But why call such a relationship, on the monetary side of things, a trust at all?

Del Mar provides a significant clue in a statement that, on first glance, seems to evidence nothing more than a difference in monetary and metals policies on the part of Occidental and Oriental societies. He states that "The governments of Persia, Assyria, Egypt, Greece and Rome made a profit on the coinage by raising the value of gold, while those of India, China, and perhaps also Japan, made their profit by maintaining, or enhancing, the value of silver."[6] In other words, for the societies of the Occident — Egypt, Assyria (and presumably Babylon), Persia, Greece, and Rome — artificially defined gold as being the metal of highest value in terms of its convertibility into more units of other metals, while, conversely, the governments of the Orient — India and China — pursued the reverse policy, of making silver the highest valued metal in terms of its convertibility into other metals. Thus, trade could be carried out between these two disparate parts of the world; the policies were in a certain sense an inevitable consequence of that trade. However, a closer examination reveals a hidden player, for such trade will inevitably create the rise of an international trading class, one which, moreover, will create its wealth precisely by trade in these precious metals, metals that are easier to transport than finished goods, and which can be exchanged in any place for such goods. In short, what is being created, from earliest times, is an international financial class of "bullion brokers," or as we would call them now, bankers. A significant question now occurs: Is it possible that, rather than such a class having emerged as a consequence of such governmental policies and trade, that the converse is true? Is it possible that there existed such a class of "international bullion brokers" who *created* these policies in various parts of the world, policies which would enhance their own power and wealth? If so, then how did they achieve and orchestrate this?

The answer to that question will consume us in the remaining chapters of this book, but in order to answer it, we must once again turn to the unusual fixation of the value of gold and silver in the Occidental cultures: Greece, Egypt, the Mesopotamian civilizations, and Rome. Why was gold — which was much easier and less costly to mine than silver — valued more highly than silver?

Del Mar has already suggested the answer; it was because in some measure, gold was regarded as *sacred*, as being under the special jurisdiction of the gods:

---

6   Alexander Del Mar, *History of Monetary Systems,* p. 89.

> The sacerdotal character conferred upon gold, or the coinage of gold, was not a novelty of the Julian constitution (of Rome); rather was it *an ancient myth put to new political use*...A similar belief is to be noticed among the ancient Greeks, whose coinages, except during the republican era, were conducted in the temples and under the supervision of priests. Upon these issues were stamped the symbolism and religion of the State, and as only the priesthood could correctly illustrate these mysteries of their own creation, the coinage — at least that of the more precious pieces — naturally became a prerogative of their order.[7]

But this raises as many questions as it resolves. For example, Del Mar has clearly indicated that the prerogative of the issuance of money and the regulation of its value was seen by ancient societies to reside clearly with the state or crown. Yet, he has now admitted that insofar as one precious metal was concerned, the issuance of money in connection with it was less a prerogative of the state or crown than it was of a religious monopoly.

Additionally, it has already been suggested that the differences in monetary policies between the Occident and Orient could serve in the long run not only to contribute to the rise of an "international class of bullion brokers," but also that the creation of such policies might even be due to the preexistence of that class, its international extent, and its ability to manipulate the respective policies of Occidental and Oriental governments. And insofar as that ability to manipulate their respective policies is concerned, when it comes to the high sacredness of gold, at least within Occidental cultures, this implies also the ability to manipulate their religions. *Succinctly stated: the bullion trust and the temple are at the minimum allies, and at the maximum, the one has infiltrated and taken over the other.* Or in Del Mar's apt observation, one now seen to be pregnant with meaning and significance just beneath its surface, the sacredness of gold "was an ancient myth put to new political use."[8]

b. A Fascinating Tangent: Byzantium, Religion, and the Money Power

A fascinating glimpse into the strength of this association between the money-issuing prerogative of the state or crown on the one hand, and religion on the other, is afforded by the Eastern Roman or Byzantine Empire, whose power and influence throughout Christian Europe during the Middle Ages extended far beyond its ever-shifting borders. That power and influence was

---

7 Alexander Del Mar, *History of Monetary Systems*, pp. 80, 81, emphasis added.
8 Alexander Del Mar, *History of Monetary Systems*, p. 80.

due precisely because of the strength of this relationship, and its hold over the cultural imagination of the Christian Middle Ages, because for all their claimed "divine prerogatives," not even the Roman popes understood their power to include the ability to make and issue money and regulate its value. That prerogative was understood to reside still with the Roman emperor in Constantinople. Noting that the very "moment when these people became Christians, or were conquered or brought under the control of the Roman hierarchy, their gold mines began to be abandoned and closed."[9] Various numismatic scholars had long proffered various unsatisfactory explanations. But, states Del Mar,

> All such futile explanations are effectually answered by the common use of Byzantine gold coins throughout Christendom. In England, for example, the exchequer rolls relating to the mediaeval ages, collated by Madox, prove that payments in gold besants[10] were made every day, and that gold coins, as compared with silver ones, were as common then as now. If metal had been wanted for making English gold coins, it was to be had in sufficiency and at once. All that was necessary was to throw the besants into the English melting-pot. As for the feeble suggestion that for five hundred years no Christian princes wished to coin gold so long as the Basileus[11] was willing to coin for them, when the coinage of gold was the universally recognized mark of sovereignty, and when, also, the profit… was one hundred per cent, it is scarcely worth answering.…
>
> The true reason why gold money was always used but never coined by the princes of the mediaeval empire relates… to that hierarchical constitution of pagan Rome, which afterwards with modifications became the constitution of Christian Rome. Under this constitution, and from the epoch of Julius to that of Alexis, the mining and coinage of gold was a prerogative attached to the office of the sovereign-pontiff, and was, therefore, an article of the Roman constitution and of the Roman religion.[12]

---

9 Ibid., p. 72.
10 Besant: the unit of money of the Eastern Roman or Byzantine Empire.
11 Basileus: from the Greek βασιλευς, meaning "king." The official title of the Emperor in Constantinople was βασιλευς Ρομαιων or "King/Emperor of the Romans." In earlier documents, the term βασιλευς Ρομαιων was used to translate the Latin term Imperator Romanorum, Emperor of the Romans. While Byzantine imperial constitutional practice could and often did acknowledge "co-emperors," or Caesars (καισηρ or "kaisers"), the term "emperor of the Romans" was exclusively reserved for the sovereign resident *in Constantinople*. For the implications of this little-noticed fact for the proper interpretation of events in mediaeval history, see Fr. John Romanides, *Franks, Feudalism, and Doctrine*.
12 Alexander Del Mar, *History of Monetary Systems*, pp. 72–73.

Thus, as Del Mar notes, prior to the Latin Crusade which attacked and successfully occupied Constantinople in 1204, no Christian prince in Europe dared coin his own gold coins prior to that event, but after it, all of them did.[13] And thus, the hidden motivation for the attack of the Christian West on the Christian East in 1204 is revealed: it was to acquire the legal authority, under the Roman constitution, to make and coin gold money. Thereafter, the money-issuing prerogative devolved to the lesser crowned heads of Europe. Prior to that point, however, it would have been sacrilege to give currency to any other (gold coinage); hence no other Christian prince, not even the pope of Rome, nor the sovereign of the Western or Mediaeval 'empire,' attempted to coin gold while the ancient Empire survived."[14] Such monies were, quite literally, viewed as "heretical money."[15]

Hence, we have yet another *possible* explanation for the preoccupation of the crowned heads of mediaeval Western Europe with alchemy during the Middle Ages.[16] Alchemy's claim was, of course, to be able to take base metals and transform them into gold. Thus, by taking imitation coins of the gold besant, coins that could not be deemed counterfeit, since they were not circulated nor coined by anyone in gold, and transmuting them into gold, an alchemical end run around Byzantium's money monopoly could be achieved.

2. The Egyptian Model: Mining, Slavery, Mercenaries, and Implications
a. Nubia and Egypt

The connections thus far exhibited between the money-coining prerogative of the state, the international "bullion brokers' trust," the temple, and now alchemy, leads one inexorably further back into history, and to the even tighter relationships between them that are exhibited in Egypt. It is here that one begins to see more clearly the sinister outlines of that international bullion brokers' trust emerging, and of its deep relationship not only to the temple, but to the deeper physics that the temple, and money, portended. Egypt is also significant for another reason, as we shall see, in that it is a symbol of the relationship of the bullion trust to the temple, a relationship that also occurs in other states and civilizations of the period.

Egypt was probably the largest gold-mining and producing state in the ancient world. The reason is simple: the Nile. As Alexander Del Mar puts it,

---

13 Ibid., p. 70.
14 Alexander Del Mar, *History of Monetary Systems*, p. 75.
15 Ibid.
16 For more on this subject, see my *The Philosophers' Stone* (Feral House, 2009), chapters one and two.

"Gold has been found in nearly every region tributary to the Nile, from the Equator to the First Cataract."[17] But no region was more connected to gold mining than was the region of Nubia, bordering modern-day southern Egypt and the Sudan. The term Nubia itself "appears to have originated in Egypt, where *Nob* or *Nub* signifies gold, hence Nubia, the land of gold."[18] Beneath the foothills of Nubia lies a vast gravel-and-sand expanse of desert, washed during the flood season by numerous streams and gullies. This region is known as the Bisharee or Bishara, the Great Nubian Desert.[19] Del Mar notes that

> Next to the mines of the Altai mountains of India, the Bisharee mines of Egypt are probably the oldest in the world; and in view of the Indian origin of the Egyptians, and the distant researches and conquests which have been made by the leading nations for the acquisition of gold, it seems not at all improbable that there existed a close connection between the discovery of the Egyptian mines and the original settlement of the country by Asiatic races.[20]

So rich were the Bisharee mines in antiquity that it is worth having a closer look at them, for it is in that closer look that we shall discern a pattern beginning to emerge.

b. Quartz, Gold, and Slavery

The Bisharee mines, as noted above, are among the oldest known in antiquity, and some of the most extensive. Additionally, the fields, like many gold mines, are also known for their abundant deposits of quartz:

> If conjecture be admitted where dates are… confused, it appears likely that the Bisharee mines were worked as early as the era of Menes, which is variously assigned to from twenty-nine to thirty-nine centuries B.C.; for in the time of that monarch or lawgiver, the Nile was

---

17 Alexander Del Mar, *A History of Money in Ancient Countries from the Earliest Times to the Present* (Kessinger Publications, reprint of the George Bell and Sons edition, London, 1885), p. 133.
18 Alexander Del Mar, *A History of Money in Ancient Countries from the Earliest Times to the Present*, p. 131.
19 Ibid., pp. 131–132.
20 Ibid., p. 138. Del Mar's mention of Egypt's possible derivation from India merits some comment of its own. This forms, as far as Del Mar is concerned, a component of the vast migrations of the Aryan peoples from the subcontinent westward into the Middle East and later Europe. But the motivation that Del Mar gives for this migration-expansion is rather different than most: "At a very remote date we find bodies of men — Lydians, Phrygians, Phoenicians, Greeks, and others belonging to Aryan races — pushing out in all directions into the desert wilds of aboriginal Europe in search of gold, silver, and copper." (p. 126). In other words, the migration was driven by monetary considerations.

diked, and from the character of this river and its surroundings there could have been no necessity to dike it until mining had surcharged its waters with sediment.

The supposed antiquity of these mines derives support from the fact that even at the period of Menes gold had been used for money in India; that Menes was an Indian conqueror and lawgiver; that the Indian Code of Manou, still extant, and assigned variously to from the fifteenth to the thirty-first century B.C., is evidently re-compiled from a much-older code, now lost; that commerce between India and Egypt existed from the remotest times mentioned in history or derived from archaeological remains; that an Egyptian expedition to India is attributed to Sesostris, B.C. 2000, and from other considerations.

However this may be, the Bisharee mines are known to have been worked for quartz so long ago as the twelfth dynasty, which Lepsius assigns to the period B.C. 2830. *From the fact, understood by every miner, that quartz is never worked so long as the placers contain the smallest practical quantity of metal, and judging from the experience of Italy, Spain, and Brazil — where extensive placer deposits were worked, as in Egypt, by the hand labour of slaves — the Bisharee places mines were at least two hundred years old when the quartz was worked under the twelfth dynasty of the Pharaohs.* [21]

Note the presence of two new factors that will loom larger and larger as we proceed: first, the presence of precious or semi-precious stones, in this case, quartz, and the working of such mines by *slaves*.

c. Diodorus Siculus and the Bishara Mines

It is the presence of slaves at the Bisharee mines that affords us our first significant clue into the minds and mentality of the ancient bullion brokers, and why they and their activities were so often associated with temples. It requires a closer look, which is afforded by the classical writer Diodorus Siculus. Del Mar notes that Diodorus visited the mines in 50 B.C. After the visit, he recorded his impressions of the mines:

On the confines of Egypt and the neighbouring countries there are regions full of gold mines, whence, with the costs and pains of many labourers, much gold is dug. The soil is naturally black, but in the body of the earth there are many veins of shining white quartz, glit-

---

21 Alexander Del Mar, *A History of Money in Ancient Countries from the Earliest Times to the Present*, pp. 138–139, emphasis added.

tering will all sorts of bright metals, out of which those appointed to be overseers cause the gold to be dug by the labour of a vast multitude of people. For the kings of Egypt condemn to these mines not only notorious criminals, captives taken in war, persons accused of false dealings, and those with whom the kind is offended, but also all the kindred and relatives of the latter. These are sent to this work, either as a punishment, or that the profit and gain of the king may be increased by their labours.

There are thus infinite numbers thrown into these mines, all bound in fetters, kept at work night and day, and so strictly surrounded that there is no possibility of their effecting an escape. *They are guarded by mercenary soldiers of various barbarous nations, whose language is foreign to them and to each other, so that there are no means either of forming conspiracies or of corrupting those who are set to watch them.* They are kept to incessant work by the rod of the overseer, who often lashes them severely. Not the least care is taken of the bodies of these poor creatures; they have not a rag to cover their nakedness; and whoever sees them must compassionate their melancholy and deplorable condition, for though they may be sick or maimed or lame, no rest nor any intermission of labour is allowed them. Neither the weakness of old age, nor the infirmities of females, excuse any from the work, to which all are driven by blows and cudgels; until, borne down by the intolerable weight of their misery, many fall dead in the midst of their insufferable labours. Deprived of all hope, these miserable creatures expect each day to be worse than the last, and long for death to end their griefs.[22]

While the passage is one to move any decent human being to compassion and pity for the poor wretches condemned to work the mines, there are two things that should be noted.

First, given the antiquity of the mines of Bisharee and the stability of Egyptian society and culture, it is reasonable to assume that the conditions at the mines had not changed throughout the several centuries that they were worked. Thus, secondly, the presence of an international contingent of mercenary guards of the mines might well be a practice dating back for centuries. The reasons for such an international contingent Diodorus makes clear: it was to deny the condemned any opportunity of bribing a sufficient contingent of guards to effect an escape.

---

22 Diodorus Siculus, *The Historical Library of Diodorus the Sicilian,* trans. G. Booth, London, 1814, cited in Del Mar, *A History of Money in Ancient Countries from the Earliest Times to the Present,* p. 141, emphasis added.

But mercenaries might signal something else, and that is a certain collusion between the Egyptian kings and those able to *supply* such a large and disparate group of mercenaries. To put it succinctly, the mercenary guard contingent subtly implies the existence of an international money power. But implication is not yet proof.

d. Money, Kings, Temples and Counterfeiters

There is, however, another indication that something more than mere coincidence is involved in the use of mercenaries at the Bisharee mines. Another clue lies in the widespread association of ancient temples with the issuance of precious metal monies:

> The archaic Chinese and Indian, as well as the early Greek coins, were often marked with emblems, which in the former cases are supposed to be, and in the latter case are known to be, religious. The mints were in the temples, and the priests monopolized, or tried to monopolize, the secrets of metallurgy. This custom may have arisen either from the cupidity of the priesthood to reap the profits of coinage, or solicitude on the part of the sovereign to prevent counterfeiting or to render it the more heinous.[23]

This fact, plus the fact of international contingents of mercenaries at Bisharee, has now amassed a significant series of questions:

1) Why were ancient mints in *temples*, especially if, as Del Mar has conceded and as we have already examined, the prerogative of money issuance and value regulation was solely a state or crown prerogative? Part of the answer, of course, lies in the fact that in most ancient cultures, particularly those of the Middle East, the King was in fact also the chief or high priest, and the priesthood itself often fulfilled the function of the state bureaucracy. Nonetheless the question remains, why associate money issuance with temples?
2) The widespread *practice* of this association in such disparate cultures as China, India, and Greece, itself must be explained. Does it lie, once again, merely in the close association of the crown prerogative with the priesthood in ancient societies, or is there something else involved, something perhaps signaled by the presence of international mercenary contingents at Bisharee?

---

23 Alexander Del Mar, *A History of Money in Ancient Countries from the Earliest Times to the Present*, pp. 11–12.

We have already suggested that the peculiar policies of Rome vis-à-vis the Orient in the regulation of gold and silver valuation suggested a certain amount of collusion between the states so affected, a collusion suggesting an international coordination of such policies. Now, we find at the Bisharee mines yet another suggestion of such international collusion in the presence of mercenaries from various cultures. The question is, then, who is doing it, and why?

Part of the answer is suggested by a passing observation made by Del Mar:

> The Egyptians... possessing the most extensive and productive gold mines which are known to have been worked at any certain period of antiquity...*furnished to other nations, including the Indians, the material out of which they too could make money.* [24]

This is the surest testament of the fact that we are now in the presence of an international bullion brokers' conspiracy, for why would India need Egyptian gold to issue money? Why could the Indian kings and princes not have issued some other money based on some other commodity which it had in abundance, and require its acceptance by law? Such was their prerogative. So why rely on imports of gold? The argument that an internationally recognized medium of exchange was needed falls to the ground, for as already noted, India and China preferred during the Roman era to value silver higher than gold, while the Occident preferred the reverse. Moreover, by diplomatic agreement and treaty, other mechanisms of exchange could have been agreed upon by various heads of state.

In any case, we now have assembled three discrete types of evidence that point to the possible existence of an international money power, or "bullion brokers" as we have called them, in ancient times:

1) The unusual bullion policies of Rome vis-à-vis the Orient, which policies were virtual opposites of each other with respect to the relative valuation of silver and gold bullion. Such reciprocal policies could, of course, highly benefit an international money power with bases of operations in both places, for by transferring gold to the east and converting it to silver, and reaping the fees thereof, and then silver to the west and converting again to gold, vast profits could be accumulated;
2) The almost universal association of temples with minting, from the Middle East to China and India; and,

---

24 Alexander Del Mar, *A History of Money in Ancient Countries from the Earliest Times to the Present*, p. 143, emphasis added.

3) The presence of an international contingent of mercenaries from various countries and states at one of antiquity's largest gold mines at Bisharee, mines which in turn are worked by slaves and prisoners of war.

There is, however, a distinctly *different* association of temples in ancient times, besides that of minting and money issuance...

B. Temples and Templates: Astronomy, Astrology, and the Alchemy of Money
1. The Temples and the Stars

...that association is with time and its measurement, in short, with astrology and astronomy.

Today, of course, it is widely accepted that ancient temples and megalithic sites were oriented to this or that astronomical event or alignment: the solstices, the precession of the equinoxes, and so on. Societies and civilizations such as Sumer and Babylonia were well-known for their strong astronomical and astrological skills, having left literally thousands of clay tablets recording their observations. But until comparatively recently in the nineteenth century, Egypt was the exception to this rule, until a British astronomer, J. Norman Lockyear, demonstrated that the same held true for Egypt's temples as well, and from the most ancient times. Lockyear's book *The Dawn of Astronomy: A Study of Temple Worship and Mythology of the Ancient Egyptians* began a cycle that continues to this day, with alternative Egyptologists such as Andrew Collins, Graham Hancock and Robert Bauval demonstrating the astronomical layouts and alignments of the site most people associate with ancient Egypt: the pyramids and temples of Giza. One might thus well view Lockyear as one of the first modern "paleophysicists," i.e., a modern scientist looking at ancient myths seriously, and attempting to reconstruct from them an underlying scientific basis.[25]

Lockyear notes that ancient civilizations' astronomical and astrological records go back almost to the very edges of known antiquity:

> We go back in Egypt for a period, as estimated by various authors, of something like 6,000 or 7,000 years. In Babylonia inscribed tablets

---

25 I coined the term "paleophysics" in the books of my *Giza Death Star* trilogy to refer to this effort to examine ancient texts and monuments for their modern-day scientific analogues in an attempt to reconstruct aspects of a much more sophisticated lost physics. This examination emerged as a methodological assumption from the supposition that there once existed a very sophisticated Very High Civilization predating ancient classical civilizations in Egypt, Mesopotamia, the Indus Valley, and China, civilizations that were declined legacies of that precursor. In Lockyear's hands no such assumption is made, so his effort does not attempt a reconstruction of that lost science.

carry us into the dim past for a period of certainly 5,000 years; but *the so-called "omen" tablets indicate that observations of eclipses and other astronomical phenomena had been made for some thousands of years before this period.* In China and India we go back as certainly to more than 4,000 years ago.²⁶

Indeed, in Babylon's case, if one is to take the assertions of the "omen tablets" seriously, these observations comprised a database assiduously compiled for hundreds of thousands of years!²⁷

Decoding the association of various Egyptian temples to various gods, and the association of the latter in turn with various celestial bodies, Lockyear states his main thesis toward the end of his book, calling it his "working hypothesis":

1. The first civilisation as yet glimpsed, so far as temple building goes, in *Northern* Egypt, represented by that at Annu or Heliopolis, was a civilisation with a non-equinoctial solar worship, combined with the cult of a northern star.
2. Memphis (possibly also Sais, Bubastis, Tanis and other cities with east and west walls) and the great pyramids were built by a new invading race, representing an advance in astronomical thought. The northern stars were worshipped possibly on the meridian, and a star rising in the east was worshipped at each equinox.
3. The subsequent blank in Egyptian history was associated with conflicts between these and other races; which were ended by the victory of the representatives of the old worship of Annu, reinforced from the south, as if north-star and south-star cults had combined against the equinoctial cult.

    After these conflicts, east and west pyramid building practically ceased, Memphis takes second place, and Thebes, a southern Annu, so far as the form of solar worship and the cult of Sit are concerned, comes upon the scene as the seat of the twelfth dynasty.
4. The subsequent historical events were largely due to conflicts with intruding races from the north-east. The intruders

---

26   J. Norman Lockyear, *The Dawn of Astronomy: A Study of the Temple Worship and Mythology of the Ancient Egyptians* (Dover, 2006), p. 2, emphasis added. I have noted elsewhere the antiquity of astronomical observations in Egypt and Babylon, q.v. my *The Cosmic War: Interplanetary Warfare, Modern Physics, and Ancient Texts* (Adventures Unlimited Press, 2007), pp. 241–243.

27   See my *The Cosmic War*, p. 241.

established themselves in cities with east and west walls, and were on each occasion driven out by solstitial worshippers who founded dynasties (eighteenth and twenty-fifth) at Thebes.[28]

We shall not concern ourselves with the lengthy arguments that Lockyear presented in order to arrive at these conclusions, but note only that the various conflicts in Egypt — given Lockyear's astronomical premise for temple orientation — were in a certain sense as much about physics as they were about religion. Given the fact that Babylonian temples were well known to be associated with various deities and celestial bodies, and that Egypt's demonstrably were as well, as Lockyear argued, *we now note the peculiar fact that temples, for a reason as yet unknown, were closely associated both with celestial alignments and the minting and issuance of bullion-based monies.*

One possible reason for the association, however, is suggested by well-known space anomalies researcher Richard C. Hoagland. We may approach Hoagland's remarks by noting a curious fact observed by Lockyear: "…In Lower Egypt the temples are pointed to rising stars near the north point of the horizon or setting north of west. In Upper Egypt we deal chiefly with temples directed to stars rising in the south-east or setting low in the south-west."[29] The fact that so many ancient temples show evidence both of a profound association with the stars through their astrological alignments, and also of international banking through the prominent association of moneychangers with those temples, is an indicator that at a deep and profound level — perhaps as a legacy of the Very High Civilization from which they sprang — these classical civilizations through their marriage of banking and astrology preserved the dim memory of a lost science that unified physics, economics, and finance.

Hoagland made the following interesting series of observations in 1992 during his presentation of the Mars anomalies evidence during a briefing at the United Nations:

> Or take the Sun itself. For the last several years, scientists have been looking for nuclear particles supposedly coming from the center of the Sun because of its thermonuclear reactions, a kind of 'chained H-bomb' model. But for the last 20 years, they haven't found anywhere near as many of these particles as models would predict; in fact, the Soviets and the Japanese, who have recently brought on line experiments to measure fundamental particles, have found none. Is

---

28 Lockyear, *The Dawn of Astronomy*, pp. 330–331.
29 J. Norman Lockyear, *The Dawn of Astronomy: A Study of Temple Worship and Mythology of the Ancient Egyptians* (Mineola, New York: Dover Publications, Inc., 2006), p. 341.

it possible that at the very center of our solar system, this incredible blazing ball of gas, around which all the worlds in this system orbit, is in fact fueled from another source? ...Well, this opens up remarkable possibilities which, when we look back at our cultures on this planet, and their attention to the Sun, others may have also noticed.

When we look at the stars, at the sky at night, we're not seeing chained and imprisoned H-bombs, we're seeing *portals* to another dimension. And the portals are glowing windows through which we can peer and glimpse the fragments of a physics from another side. What is stunning is when you take that metaphor, and you go back and you read the actual Egyptian descriptions in the hieroglyphs for Sirius, the brightest star in the sky, the description is (that) Sirius is a doorway. Now what did they know? What did they know?

Now, looking at that as a twentieth-century Egyptologist, or as a modern anthropologist, or as any of the academics looking at the history of man and archaeology, the obvious answer is: 'They didn't know anything. It was all primitive superstition.' ... (But) we can now buttress with specific mathematical linkages, which indicate strongly (that) the Egyptians, among others, were profoundly aware that the night sky and all of reality is somehow circumscribed by the physics behind the circumscribed geometry given to us now by the Monuments of Mars. And that of course is an amazing statement to be able to make.

But there's another, because the thought occurred then, after we had figured this part out — the potential, the possibility — that if in fact the Sun is a 'gate' to another dimension, and (that) the energy we see is merely a lower-level transformation of a higher-level force or energy or process, it may be possible to harness such a process on earth and to create an actual hyper-dimensional technology. What could one do....?[30]

What, indeed, did ancient Egypt, and by implication, all other ancient cultures, with their persistent and consistent association of stars with deities or "higher intelligences" and other dimensions, know? Was there indeed a deeper physics behind their astrological obsessions? And what *could* one do with that physics? Hoagland has also suggested something else as well, and that is that there is a physics connection with the ancient fixation on "sacred geometry" and the positioning of various temples and buildings on sites that embody such geometries, in this case, on Mars. And why does one find *moneychangers* invariably associated with these temples? One answer is hinted at by Lockyear.

---

30 Richard C. Hoagland, DVD of *Hoagland's Mars: Vol. 2: The United Nations Briefing:* (UFO TV), at 37:13–40:01 in the presentation. Hoagland made his remarks at the U.N. in February 1992, emphasis added.

2. The Peculiar Resemblance of Pantheons Between Cultures

Lockyear makes yet another pertinent observation relevant to our purposes. To understand its significance, one need only consider the fact that if temples are associated with various gods — whether in Egypt or Mesopotamia — and various gods in turn are associated with various celestial bodies, then there should be considerable parallel between the pantheons of various civilizations, since, in a word, the underlying astronomical data are the same. This, he observes, is in fact what one actually sees. There is a peculiar parallel between the Egypt god Annu or An, and the Babylonian Anu.[31] More to the point, on an astrological-astronomical reading, the gods blend into each other. For example, the chief god of the Mesopotamian city of Eridu was Ea, a.k.a. Enki, a.k.a. Oannes, "symbolized as a goat-fish."[32] This Ea or Enki sired a son, Tammuz, who was "in some way associated with Asari," who, notes Lockyear, has a name suspiciously similar to the Egyptian Osiris.[33] The god Tammuz in turn ultimately becomes the notoriously murderous god Nergal in Chaldaea.[34] Nergal by similar astronomical conversions eventually actually becomes "the Spring Sun Marduk" of Babylon itself.[35] In other words, lurking behind all these transformations and pantheonic parallels is the association of Enki with Nergal and Marduk, and via Tammuz, with the Egyptian Osiris. Such pantheonic parallels are, once again, inevitable if, as Lockyear shows, the gods themselves are associated with various celestial bodies. The patterns of the heavens, as it were, formed a template with which to associate certain gods and their activities.

However, given the fact that we now have the close association of temples not only with astronomy and astrology, but with money issuance and therefore with mining, and given the fact that we have discovered three significant indicators that we may also be looking at the discrete and subtle actions of an ancient international "bullion brokers" power, the similarity of pantheons may have other than just astronomical causes. It may indeed be a product of the deliberate manipulation of those pantheons by an entity that has every appearance of being international in extent and influence in ancient times. But to what purpose? The answer to that question must unfortunately await the next chapter, as there is a further association of gods, gold and temples to be explored:

---

31 Lockyear, *The Dawn of Astronomy*, p. 363.
32 Ibid., p. 372.
33 Ibid.
34 Ibid.
35 Ibid., p. 373.

3. Gold, Gods, and Gems

That other association is with precious and semi-precious gems. It is a peculiar fact, notes researcher George Frederick Kunz, that

> From the earliest times in man's history gems and precious stones have been held in great esteem. They have been found in the monuments of prehistoric peoples, and not alone the civilization of the Pharaohs, of the Incas, or of the Montezumas invested these brilliant things from Nature's jewel casket with a significance *beyond the mere suggestion of their intrinsic properties.* [36]

But why give such items a significance beyond "their intrinsic properties"? One answer might lie in their well-known association with astrology from ancient times:

> The magi, the wise men, the seers, the astrologers of the ages gone by found much in the matter of gems that we have nearly come to forgetting. With them each gem possessed certain planetary attractions peculiar to itself, certain affinities with the various virtues, and a zodiacal concordance with the seasons of the year. Moreover, these early sages were firm believers in the influence of gems in one's nativity — that the evil in the world could be kept from contaminating a child properly protected by wearing the appropriate talismanic, natal, and zodiacal gems.[37]

That is to say, gems, since they were associated with certain celestial alignments, planetary bodies, and houses of the zodiac, could quite literally, on the ancient view, draw down (or ward off!) the influences of those bodies and alignments. But why did such a belief originate? What is its underlying cause?

Search as one might for the answer to those questions amid ancient texts and tomes,[38] one always comes up empty-handed, being met with the bare and naked assertion *that* it is so, with perhaps now and then an attempt — usually unconvincing — as to *why* it is so. Kunz even notes that during the Renaissance "an effort was made to find a reason of some sort for the

---

36 George Frederick Kunz, *The Curious Lore of Precious Stones* (Dover, 1971), p. 1, emphasis added.
37 Ibid.
38 Kunz draws upon several sources, from ancient and Hindu (*The Curious Lore of Precious Stones*, pp. 13–14), to mediaeval (pp. 14–15), Hermetic (p. 16), and even Christian patristic sources (p. 16).

traditional beliefs."³⁹ Nor is he above suggesting that eventually, science itself might suggest the reasons, or rather, re-discover them.⁴⁰

a. The Traditional Powers Associated with Gems

Leaving aside the speculative scientific case for this association to a later chapter, we shall catalogue here but a few of their alleged powers and virtues attributed to them by ancient lore. One almost universal property attributed to them is the ability to heal various maladies and infirmities.⁴¹ Oftentimes these were associated specifically to their various colors and relative luminescence.⁴² In one case, a Hindu tradition even describes diamonds as "Indra's weapon."⁴³ There are, in the Sumerian tradition, stones of "love" and "hate," or, as the Sumerian literally has it, "un-love."⁴⁴ And as readers of my book *The Cosmic War* know, one Sumerian epic, *The Epic of Ninurta*, is nothing but a rather dull inventory-taking of some extraordinary stones that were captured after a gruesome war, shades of the imaginary scenario with which we began this book!⁴⁵

And there are even some rather special cases of powers attributed to gemstones that merit closer attention.

(1) Special Case Number One: Invisibility

Kuns cites a passage from the 1659 treastise "The Faithful Lapidary" by Thomas Nicols that is worth mentioning here. Nicols begins by enumerating a typical list of the reputed and assumed powers of precious gems:

> *Perfectionem effectus contineri in causa.* But it cannot truly be so spoken of gemms [sic] and pretious [sic] stones, the effects of which, by Lapirists are said to be, the making of men rich and eloquent, to preserve men from thunder and lightning, from plagues and diseases, to move dreams, to procure sleep, to foretell things to come, to make men wise, to strengthen memory, to procure honours, to hinder fascinations and witchcrafts, to hinder slothfulness, to put courage into men, to keep men chaste, to increase friendship, to hinder difference and dissention…

---

39 Ibid, pp. 1–2.
40 Ibid., p. 2.
41 Ibid., pp. 6, 28.
42 Ibid., p. 28.
43 Ibid., p. 343.
44 Ibid., p. 35.
45 Farrell, *The Cosmic War,* pp. 204–233.

This is a fairly standard list, and one can even understand the attribution of the power "to foretell things to come" to stones, for given their astrological associations, it stands to reason that the predictive properties of astrology should be associated with gemstones. But then comes a rather unusual statement:

> ... and to make men invisible, as is ... affirmed by Albertus and others... and many other strange things are affirmed of them and ascribed to them, which are contrary to the natures of gemms. [46]

One wonders just what other "strange things" could possibly be attributed to the powers of gemstones beyond conferring invisibility!

The belief that some stones could confer invisibility was echoed by a fourteenth-century alchemist, Pierre de Boniface, who claimed that diamonds could make the wearer invisible.[47] Given that an alchemist is making these claims, one might be justified in drawing the speculative conclusion that this operation could only occur at certain times and under certain conditions, given the well-established fact that most alchemists claimed their operations had to be performed at certain times in order for them to work.

(2) Special Case Number Two:
Magnetic Levitation, or Anti-Gravity

An even more stunning claim is made in a Viennese print of 1709 and an accompanying text of Valentini, "Museum museorum oder die vollständige Schau-Bühne," an obscure manuscript published in Frankfurt-am-Main in 1714:

> According to the text accompanying a curious print published in Vienna in 1709, the attractive qualities of the so-called coral-agate were to be utilized in an air ship, the invention of a Brazilian priest. Over the head of the aviator, as he sat in the air-ship, there was a network of iron to which large coral-agates were attached. *These were expected to help in drawing up the ship, when, through the heat of the sun's rays, the stones had acquired magnetic power.* The main lifting force was provided by powerful magnets enclosed in two metal spheres; how the magnets themselves were to be raised is not explained.[48]

---

46 Kunz, *The Curious Lore of Precious Stones*, p. 7, emphasis added.
47 Ibid., p. 72. Additionally, Kunz notes that Arab and Persian authors believed diamonds could also confer invincibility.
48 Kunz, *The Curious Lore of Precious Stones*, pp. 52–53.

Here it is evident that, for whatever obscure reason, coral-agate was thought to somehow amplify the effects of magnets in conjunction with exposure to the rays of the sun!

*(3) Special Case Number Three:
Light-Absorbing and Emitting Stones*

A much more serious special case is mentioned by Kunz in connection with the peculiar properties of Brazilian diamonds:

> The power of absorbing sunlight or artificial light and then giving it off in the dark is only possessed by certain diamonds. These are Brazilian stones, slightly milky in tint, or blue-white as they are often termed, and it is an included substance and not the diamond itself giving it out. Willemite, kunzite, sphalerite (sulphide of zinc) and some other minerals possess the same power. Their peculiar property may be due to the presence of a slight quantity of manganese or to that of some of the uranium salts. That it is only the ultra-violet rays that are thus absorbed by these diamonds is proved by the fact that the phenomenon is not observable when a thin plate of glass is interposed between the sunlight of artificial light and the diamond, as glass is not traversed by these rays….
>
> On the other hand all diamonds phosphoresce when exposed to the rays of radium, polonium, or actinium, even when glass is interposed. Treating of some of the aspects of phosphorescence in diamonds, Sir William Crookes says:
>
> "*In a vacuum, exposed to a high-tension current of electricity, diamonds phosphoresce of different colours,* most South African diamonds shining with a bluish light. Diamonds from other localities emit bright blue, apricot, pale blue, red, yellowish-green, orange, and pale green light. The most phosphorescent diamonds are those which are fluorescent in the sun. One beautiful green diamond in my collection, when phosphorescing in a good vacuum, gives almost as much light as a candle, and you can easily read by its rays. But the time has hardly come when diamonds can be used as domestic illuminants!"[49]

---

49   Kunz, *The Curious Lore of Precious Stones*, pp. 171–172, emphasis added. This is a rather interesting comment by Crookes, for archaeologists have long puzzled how the ancient Egyptians managed to *see* inside their temples. If one grants the proposition that they may have known of and used electricity, then might they perhaps have discovered this quality of diamonds? Or was such knowledge another legacy bequeathed to Egypt?

This is a rather interesting comment by Crookes, for archaeologists have long puzzled how the ancient Egyptians managed to *see* inside their temples. If one grants the proposition that they may have known of and used electricity or other sources of phosphorescence, then might they perhaps have discovered this quality of diamonds or other stones?

In answer to this question, Kunz makes an interesting observation:

> An old treatise in Greek, said in its title to come from "the sanctuary of the temple," and containing material, partly of Egyptian origin, may help us to understand something of the processes employed by a temple priest to impress the common people by the sight of luminous gems. The writer of the treatise declares that for the production of "the carbuncle that shines in the night" use was made of certain parts (he says "the bile") of marine animals whose entrails, scales and bones exhibited the phenomenon of phosphorescence. If properly treated, precious stones (preferably carbuncles) would glow so brightly at night "that anyone owning such a stone could read or write by its light as well as he could by daylight."[50]

Note that the source for this assertion comes from ancient Greek alchemical texts. And alchemy is, of course, invariably associated with Egypt and its temples. We may reasonably assume, then, that such properties, if known to the ancient Egyptian temple priesthoods, were among their most closely guarded secrets.

b. Gemstones, the Zodiac, and the Hebrew High-Priest's Ephod, or Breastplate

Perhaps the most familiar association of precious gemstones with the temple and religion in Western culture is the breastplate, or *ephod,* of the Hebrew high priests described in the Old Testament of the Bible, so it is worth pausing, before concluding this chapter, to take stock of the various associations of the breastplate in Jewish lore. Kunz observes that in Rabbinical legend "it is related that four precious stones were given by God to King Solomon; one of these was the emerald. The possession of the four stones is said to have endowed the wise king with power over all creation."[51] This is a rather breathtaking claim for the power of these stones, whatever else they were in addition to the emerald.

---

50 Kunz, *The Curious Lore of Precious Stones,* pp. 173–174, citing *Collection des anciens alchemists grecs,"* ed. M. Berthelot, trans. (Paris, 1887, 1888), pp. 336–338, 351–352.

51 Ibid., p. 78, citing Weil, *Biblische Legenden,* p. 225.

In my book *The Cosmic War* I noted that similar claims were made for the Sumerian "Tablets of Destiny," stones which allegedly conferred "all the power of the universe" on their possessors, making them a technology highly prized — and fought over — by the various Sumerian gods.[52] Here we encounter a similar claim, from a much later period, and in a different culture. Solomon, of course, is known not only for his legendary wisdom and wealth, but is also associated with the building of a magnificent temple.

It is possible, if such stones existed, that the celebrated Hebrew king wore them in a royal version of the priestly breastplate or *ephod*, for there do exist Assyrian versions of the breastplate, composed of seven gems — one for each of the seven planets of Mesopotamian astrology — and to be worn by the king:

> Among the Assyrian texts giving the formulae for incantations and various magical operations, there is one which treats of an ornament composed of seven brilliant stones, to be worn on the breast of the king as an amulet; indeed, so great was the virtue of these stones that they were supposed to constitute an ornament for the gods also. The text, as rendered by Fossey, is as follows:
> "Incantation. The splendid stones! The splendid stones! The stones of abundance and joy.
> "Made resplendent for the flesh of the gods.
> "The *hulalini* stone, the *sirgarru* stone, the *hulalu* stone, the *sandu* stone, the *uknu* stone.
> "The *dushu* stone, the precious stone *elmeshu*, perfect in celestial beauty.
> "The stone of which the *pingu* is set in gold.
> "Placed upon the shining breast of the king as an ornament.
> "*Azagsud, high-priest of Bel, make them shine, make them sparkle!*
> "Let the evil one keep aloof from the dwelling!"[53]

Note carefully the reference to the Assyrian high priest, exhorting him to make the stones shine and sparkle, suggesting that not only did Egyptian temple priests have a secret of making gemstones phosphoresce, but that their Mesopotamian counterparts did as well.

Similar breastplates were associated with Babylon and even the king of Tyre as well.[54] So what of the Hebrew high priest's breastplate? Here there is a rich rabbinical, and even Muslim, tradition. The 12 stones of the high priest's

---

52 Q.v. my *The Cosmic War*, pp. 204–232.
53 Kunz, *The Curious Lore of Precious Stones*, p. 230, citing Fossey, *La Magie Assyrienne* (Paris, 1902), p. 301, emphasis added.
54 Ibid., p. 231.

breastplate are first associated with the 12 angels who "guard the gates of Paradise."[55] The Jewish historian Flavius Josephus, moreover, records a tradition in which the high priest's garments were also fastened at the shoulders with phosphorescing stones.[56] Moreover, each of the 12 stones of the breastplate was engraved with the names of each of the 12 tribes.[57] Josephus further associates the 12 stones of the *ephod* to the 12 months of the year,[58] and to the zodiac as well.[59] By mediaeval times, Jewish tradition definitely associated the 12 tribes to the signs of the zodiac in the following correspondence:

Judah	Aries
Issachar	Taurus
Zebulun	Gemini
Reuben	Cancer
Simeon	Leo
Gad	Virgo
Ephraim	Libra
Manasseh	Scorpio
Benjamin	Sagittarius
Dan	Capricorn
Naphtali	Aquarius
Asher	Pisces[60]

The breastplate, like the Ark of the Covenant, is one of those objects of historical and religious power from the Jewish temple that, after the fall of Jerusalem to the Roman armies in 70 A.D., seems to disappear from history. Where did the *ephod* with its precious gems go?

Kunz's answer is worth citing extensively:

---

55   Ibid., pp. 275–276.
56   Ibid., p. 277.
57   Kunz, *The Curious Lore of Precious Stones,* pp. 277–278.
58   Ibid., p. 309
59   Ibid., p. 310.
60   Ibid., p. 314. Kunz also notes that the wearing of a breastplate consisting of 12 ornaments is unique to the Hebrews *and to Egypt:* "That an Egyptian origin should be sought seems most probable. A breast ornament worn by the high priest of Memphis, as figured in an Egyptian relief, consists of twelve small balls, or crosses, intended to represent Egyptian hieroglyphics. As it cannot be determined that these figures were cut from precious stones, the only definite connection with the Hebrew ornament is the number of figures; this suggests, but fails to prove, a common origin. The monuments show that the high priest of Memphis wore this ornament as early as the fourth Dynasty, or, approximately 4000 B.C." (p. 282) Kunz also notes that after the Babylonian captivity a second breastplate was most likely made. (p. 280). It is possible that the Hebrew association of the 12 stones of the breastplate and of the 12 tribes to the 12 zodiacal signs received an impetus from Babylonian preoccupations with astrology and the associations of gemstones with celestial bodies and their influences.

(T)he treasures of the temple were carried off to Rome, and we learn from Josephus that the breastplate was deposited in the Temple of Concord, which had been erected by Vespasian. Here it is believed to have been at the time of the sacking of Rome by the Vandals under Genseric, in 455, although Rev. C.W. King thinks it is not improbable that Alaric, king of the Visigoths, when he sacked Rome in 410 A.D., might have secured this treasure. *However, the express statement of Procopius that "the vessels of the Jews" were carried through the streets of Constantinople, on the occasion of the Vandalic triumph of Belisarius, in 534, may be taken as a confirmation of the conjecture that the Vandals had secured possession of the breastplate and its jewels.*

It must, however, be carefully noted that Procopius nowhere mentions the breastplate and that it need not have been included among "the vessels of the Jews." It appears that this part of the spoils of Belisarius was placed by Justinian (4830565) in the sacristy of the church of St. Sophia. Some time later, the emperor is said to have heard of the saying of a certain Jew to the effect that, until the treasures of the Temple were restored to Jerusalem, they would bring misfortune upon any place where they might be kept. If this story be true, Justinian may have felt that the fate of Rome was a lesson for him, and that Constantinople must be saved from a like disaster. Moved by such considerations, he is said to have sent the "sacred vessels" to Jerusalem, and they were placed in the Church of the Holy Sepulchre.

This brings us to the last two events which can even be plausibly connected with the mystic 12 gems — namely, the capture and sack of Jerusalem by the Sassanian Persian king, Khusrau II, in 615, and the overthrow of the Sassanian Empire by the Mohammedan Arabs, and the capture and sack of Ctesiphon, in 637. If we admit that Khusrau took the sacred relics of the Temple with him to Persia, we may be reasonably sure that they were included among the spoils secured by the Arab conquerors, although King, who has ingeniously endeavored to trace out the history of the breastplate jewels after the fall of Jerusalem in 70 A.D., believes that they may be still "buried in some unknown treasure-chamber of one of the old Persian capitals."

A fact which had generally been overlooked by those who have embarked on the sea of conjecture relative to the fate of the breastplate stones is that a large Jewish contingent, numbering some twenty-six thousand men, formed part of the force with which the Sassanian Persians captured Jerusalem, and they might well lay claim to any Jewish vessels of jewels that may have been secured by the conquerors. In this case, however, it is still probable that these precious objects fell into the hands of the Mohammedans who captured Jerusalem in the same year in which they took Ctesiphon.[61]

Contrary to all those who believe that the Ark and breastplate ended up in the West in the hands of the Vandals and Visigoths, Kunz is correct, for by injecting the East Roman Empire back into the equation, he is clearly saying, "Look east rather than west" if one wishes to discover the whereabouts of the old Jewish temple treasure.[62]

C. Conclusions

The following constellation of relationships, and implications, has been exhibited in this chapter:

1) There are at least three disparate and significant clues to the existence of an international money power in ancient times, and to its reliance upon precious metal bullion as a medium of monetary exchange:
    a) The unusual bullion policies of Rome vis-à-vis the Orient, which policies were virtual opposites of each other with respect to the relative valuation of silver and gold bullion. Such reciprocal policies could, of course, highly benefit an international money power with bases of operations in both places, for by transferring gold to the east and converting it to silver, and reaping the fees thereof, and then silver to the west and converting again to gold, vast profits could be accumulated;
    b) The almost universal association of temples with minting, from the Middle East to China and India; and,
    c) The presence of an international contingent of mercenaries from various countries and states at one of antiquity's largest gold mines at Bisharee, mines which in turn are worked by slaves and prisoners of war.
2) In addition to the association of temples with the minting and issuance of money, and therefore with mining activity, these temples are also associated with astronomy and astrology;

---

61 Kunz, *The Curious Lore of Precious Stones,* pp. 283–285, emphasis added.
62 An important consideration in this regard is that the Emperor Justinian, who made it his deliberate and conscious goal to restore Roman rule over Rome itself, would most likely, through his brilliant military mastermind Belisarius, have exacted from the Vandals the return of all treasure they took from the Western Rome.

3) Additionally, astrology and astronomy are in turn associated with precious gemstones, each type of which is associated with various planetary bodies and their influences;
4) The various gods of the pantheons of the Middle East, since they can at one level be decoded as celestial bodies, are parallel, that is to say, the various pantheons themselves show remarkable parallels due to their similar astrological derivation. At this juncture, a significant question was implied:
   a) Could the parallelism of various pantheons be due not only to their astronomical derivation, but also because of the activity of that class of international bullion brokers that appears to be associated with the temple in all lands, and influencing the bullion policy to its advantage in several states?
   b) If so, then is there a deeper reason for their association with the practice of astrology in the temples? If there is, then what is it? And what is the connection between all this, and precious gemstones?
   c) These gemstones were seen in turn by the ancients and in esoteric tradition to have certain powers; among these we note the following:
      i) They possessed the power to heal;
      ii) They possessed the power to predict the future;
      iii) They could phosphoresce under certain conditions, and this was most likely a closely held secret of the temple priesthoods in various civilizations;
      iv) They conferred invisibility;
      v) They conferred the power to levitate;
      vi) In Solomon's case, certain stones conferred power over creation.

We are thus in the presence of a complex dynamic, and making sense of it will not be easy, but before we can address this issue, we must assemble more data by taking a closer look at the activities of this ancient "bullion brokers' international," for those activities will disclose the common thread and clue: alchemy.

Five

Money, Monotheism, Monarchies and Militaries
The Thesis of David Astle

∴

"They are but pudgy and sly little men as much overwhelmed by the monster they have raised, as are the foolish nations that permitted them to do so."
—David Astle[1]

A. The State of Evidence and the Need for Speculation

David Astle is a researcher who has assembled a massive database and well-argued case for the existence, in ancient times, of an international bullion brokers' trust, allied with the temple, manipulating governments, religions, pantheons and policies behind the scenes for its own benefit and agenda. As such our examination here can only be a brief review of his work. It is to be emphasized that this review is but a small sampling of the types of evidence and analyses that Astle provides.[2]

This being said, Astle himself provides a strong cautionary note on the difficult state of the evidence, and the need for speculation. Noting that all ancient societies had a system of state warehouses connected with its money coining and regulating power, Astle states that

> In almost all of the works of the great archaeologists and scholars specializing in the ancient civilizations, there is a virtual silence on that

---

1 David Astle, *The Babylonian Woe,* www.jrbooksonline.com/PDF_Books/the_babylonian_woe.pdf, p. 92.
2 I should also stress that I do not share all of Astle's theological or spiritual leanings, though these do not detract or interfere with the main argument of his book.

all important matter, the system of distribution of food surpluses, and surpluses of all those items needed towards the maintenance of a good and continuing life so far as were required by climate and custom.

In all the writings of these great and practical scholars, the workings of that mighty engine which injects the unit of exchange amongst the peoples, and without which no civilization as we know it can come to be, is only indicated by a profound silence. *Of the system of exchanges, of the unit of exchange and its issue by private individuals, as distinct from its issue by the authority of sovereign rule, on this all important matter governing in such totality the conditions of progression into the future of these peoples, not a word to speak of.*[3]

There is a related question, and Astle is quick to perceive it: not only is there a paucity of evidence relating how such an international bullion brokers' class might have arisen nor evidence as to "what created them,"[4] it is the pervasive and almost universal absence of such evidence that "borders on the mystifying." Additionally, there is in scholarly literature on ancient international monetary policy a "complete failure to speculate on those most important matters at all." But most importantly, Astle observes, there is a complete absence of speculation on "the true nature of the *energy source by which such machinery was driven.*"[5]

While Astle clearly throughout his work believes that energy source to be the widespread practice of slavery in ancient times, and while we tend to agree with him insofar as that goes, there is a *deeper* question that Astle's comments disclose: was there another and deeper energy source that existed in, or was being sought by, those ancient "bullion brokers'" fraternities besides slavery?

In any case, it is this perplexing silence on such crucial matters that calls for, indeed almost compels, speculation to fill the void.

> Practically no information seems to exist of the growth of private money creation in the days of the ancient city states of Mesopotamia of which, because of their records being preserved on fire-baked clay, more is known than of more recent civilizations; and the gap must necessarily be filled by a certain amount of speculation. *Little is known of the beginnings of the fraudulent issuance by private persons of the unit of exchange, as in opposition to the law of the gods from whom kings in ancient times claimed to derive their divine origin.*[6]

---

3   David Astle, *The Babylonian Woe*, p. 3, emphasis added.
4   Ibid.
5   Ibid., emphasis added.
6   Astle, *The Babylonian Woe*, p. 5, emphasis added.

Thus, we now encounter a new dynamic: if the assumption of the existence of a private international bullion brokers' power has merit, then there is a conflict between its private money-creating powers and that of the various states within which it exists. There are thus two operating units of exchange according to Astle: on the one hand, one has the circulation of coins of bullion as a unit of exchange, which coins were issued by temples and their allies, the bullion brokers, those who controlled the mines and slaves. On the other hand, one has the issuance — in Mesopotamia — of clay units of exchange, issued by the state power itself in conjunction with the temple, which units were backed by the surpluses of the state warehouses.[7]

1. Ancient and Modern Banking Conspiracy

Notwithstanding all this, Astle admits, largely on the types of considerations advanced in the previous chapter here, that there are significant indicators within ancient societies of

> ...(T)he existence of a far-reaching conspiracy in respect to monetary issuance influencing the progression of man's history in the earliest times of which written records exists. It is also outstandingly clear that it was parent to that acknowledged and most obvious conspiracy such as exists today.[8]

There is, in other words, more or less a continuity between ancient times and modern ones. The question is: what is the nature of that continuity? Is it merely conceptual, rooted in the similarities of goals, methods, and beliefs of the people engaged in such activites? Or is the continuity deeper? Does it extend to actual descent of certain families and groups down through history? As will be seen in this chapter, one can answer at least the first question with a resounding affirmative. And as will be seen momentarily, there are also significant indicators that the second question may be affirmatively answered in some cases as well.

B. The Medium of Exchange and Bullion as an Order on State Warehouses

Astle's argument, while profound, is not always easily seen, scattered as it is throughout his book. Care is required in noticing its steps and taking due account of them in order that its transparency may be fully exhibited. We have

---
7 Ibid.
8 Ibid., p. 10.

already noted in the previous chapter, and in this one, the following things:

1) The regulation of the value of money was established by law and regulated by the state or crown, whose sole prerogative was the issuance and regulation of such units of money;
2) There was a deep and profound relationship between the minting and issuance of such money in the form of bullion-based coins by the crown or state power, and the temple;
3) *Prior* to this period, however, in Mesopotamia, the units of exchange, as noted by Astle, were simply clay tablets of credit against the surplus in the state warehouse.

It is at this last point that the story, and Astle's argument, really begins.

The reason is easily seen, for if one were to conduct trade *between* states, the issuances of money in the form of letters of credit against a given state's warehouse surplus would inhibit, rather than aid, such trade. Another mechanism was required that would not require a diplomatic negotiation between states for each and every trade. Thus, at one and the same time as a medium of international exchange is more or less agreed upon, there also arises an international class of merchants conducting such trade and exercising sweeping influence over those units of exchange. Astle states it with his characteristic sharpness and concision:

> Through stealthy issue of precious metal commodity money into circulation amongst the peoples, replacing that money which represented the fiat of will of the god of the city and which was merely an order on the state warehouses through his scribes, this internationally-minded group from the secrecy of their chambers were able to make a mockery of the faith and belief of the simple people.[9]

In other words, at a certain juncture and by the nature of the case, there will be *two* monies in circulation, one created by a private group and based on bullion commodity money issued *as a facsimile of the state money for which it can be traded and substituted*, and the second, the original state-issued money. It should be noted that this situation is almost exactly paralleled in modern times by the issuance of state money and private notes both in England and in the United States.[10] In both the ancient and modern instances, the privately-

---

9   Astle, *The Babylonian Woe*, p. 9.
10 Consider only the fact that in the United States, since the establishment of the Federal Reserve system, there were initially several types of paper notes in circulation: the state's silver and gold certificates and United States *bills*, and the private bank's "Federal Reserve *notes*." It should also be noted that it is

issued money gradually and with inexorable inevitability replaces the state issuance, and with that occurrence, the power of the private issuers of such monies is established over a society.

But how was this substitution actually made? What was the stratagem that made it possible for ancient bullion brokers to achieve this extraordinary and sweeping influence over ancient states? The answers to these questions require an even closer look.

1. The Control of Mining and Bullion

The issuance of bullion-based monies implies something very significant, and Astle is quick to perceive it:

> The whole notion of the institution of precious metals by weight as common denominator of exchanges, internationally and nationally, cannot but have been disseminated by a conspiratorial organization fully aware of the extent of the power to which it would accede, *could it but maintain control over bullion supplies and the mining which brought them into being in the first place.*[11]

We have already seen how Egypt's Bisharee mines, with their contingent of international mercenary guards, implies the existence of such an international class of merchants and bullion brokers.

As an inevitable consequence of such activity, "the law of the ruler previously exercised towards the well-being of the people in that might live a good and honourable life accordingly became corrupted,"[12] since that crown power was inevitably in collusion with the very class undermining its power. This fact is attested by the rise of laws of punishment condemning lawbreakers to a lifetime of slavery working the semi-private mines which supplied both the state and the merchants with the medium of exchange that created their profits. Again, the situation is almost exactly paralleled by modern times, where one again encounters a private monied class in collusion with the state power for the expansion of their powers over the great masses of people.

A further implication falls out from this, and again, one perceives instantly the modern parallel:

---

now all but impossible to find any of the former notes in general circulation, since the Federal Reserve makes it a policy to pull and destroy any bills that are not the interest-bearing debt notes it does not issue privately, and since collectors, knowing this, themselves often pull such notes for their collections or to sell to other collectors.
11 Astle, *The Babylonian Woe*, p. 10, emphasis added.
12 Ibid., p. 8.

Kings largely became the mouthpiece and sword arm of those semi-secret societies that controlled the material of money as its outward and visible symbols came to be restricted to gold, silver, and copper... The fiat of the god in heaven which had been the decisive force behind that which brought about an equitable exchange, was replaced by the will of those classes controlling the undertones of civilization, leaders of the world of slave drivers, caravaneers, outcasts and criminals generally, such as was to be discerned on the edges of the ancient city civilizations...... The instrument of this will was precious metal, whose supply was controlled by the leaders of these classes through their control of the slave trade, since mining was rarely profitable in the case of the precious metals, except with slave labour...[13]

In support of this contention, Astle notes that prior to the fourth century B.C., the institution of slavery "had been more in the nature of a benign custom similar to the custom of the bonded servant or apprentice of the 18th and 19th centuries in Northern Europe."[14] After that period, however, it became an extensive and international trade, complete with whips and chains and all the imagery of human suffering one associates with the institution.

Thus we now have an increasingly expanded, and interesting, constellation of relationships:

1) A relationship between priest-kings, the temple, money minting and issuance, and a private class of international bullion brokers;
2) A relationship of the latter class with the slave trade necessary to conduct profitable mining to circulate bullion as a medium of exchange;
3) A relationship of the temple with money minting and issuance on behalf of that private class of bullion brokers and slave traders; and finally,
4) A relationship of the temple to astrology and all its associated accoutrements: gems, sacred geometry, earth grid locations for temples, priest-magicians, forecasters, and so on.

The relationship between this private class and the temples in various states exercised a correspondingly corrupting influence on their priesthoods:

...(I)t is clear that with the growth of silver in circulation between private persons, and between private persons and states, as now would become an inevitability, that which had been total economic control

---

13   Ibid.
14   Astle, *The Babylonian Woe*, p. 29.

from the gods through his servants in the Ziggurat, was bypassed, and merchants were not able to deal privately using their credit, or powers of abstract money creation. They were also able, through their control of distant mining operations, to afflict a previously dedicated priesthood with thought of personal possession, and through the control of manufacture of weapons in distant places, they were able to arm warlike peoples towards the destruction of whosoever they might choose.[15]

Notice the dynamic subtly suggested here, for through its associations with the various temple priesthoods such a class also influences war policies of the various states under its influence, and these as might be expected could often be in the guise of religious conflicts.[16]

Citing scholar T.B.L. Webster, Astle comments at length on this dynamic:

"The Alalakh tablets also record copper distributed to smiths, but not in addition it is to be used for making baskets or arrowheads; and the King of Assyria sent copper to Mari to be made into nails by the local craftsmen. A Report from Pylos that the woodcutters in two places are delivering 150 axles and 150 spars for the chariot factory may be compared with the Ugartic texts on the delivery of wood for the making of arms, and a note of wood delivered to the carpenters for the construction of wagons in Alalakh. We made here also from Pylos a list of wooden objects made, a list of vessels received by men (perhaps Mayors) in various places and a note of pieces of ivory; to set beside this rather slender evidence of Mycenaean manufacture, Alalakh provides a record of sixty-four business houses and their produce; they include smiths, leather workers, joiners, and cartwrights."[17]

Thus it seems that where the conception of money as to a silver standard existed as at Ugarit and Alalakh, so also existed organized industry, including outstandingly the private manufacture of arms under methods that appear to be those of semi-mass productions. It is not without significance that this early era of privately issued money (such as was silver money), and consequent private industry, particularly that which was devoted to arms manufacture, was in certain areas so coincidental with the massive movements of warlike peoples, and the collapse of ancient empires that had

---

15 Astle, *The Babylonian Woe*, pp. 14–15.
16 Again, it should be recalled that Lockyear proposed that the astronomical bases of various Egyptian temples could also be a basis of these religious conflicts.
17 Citing T.B.L. Webster, *From Mycenaean Homer*, p. 22.

> *lived long under the pattern of life known as that of the Ancient Orient.* Conquering peoples needed the best of arms. It seems that the best of arms were obtainable from private industry; and private industry in its turn needed silver or gold or labour, which was slaves, in payment. Both were obtainable as the result of war. Therefore parallel, though not entirely the same as in today, the more war, the more industry, and the more the need for the products of the money creators' ledgers. Hence became the more absolute the control of that which most of all designs industry and its accompanying slavery in one form or another, namely, the private money creative power.[18]

In other words, the clues have been there all along, it is just that no one has noticed them, for the great war-making empires of Mesopotamia were involved with private arms manufacture and private money issuance. This fact will loom only larger as we proceed.

But such large manufacturing and monetary concerns only lead inexorably back to the relationship between crown and temple. The renowned historian Christopher Dawson outlines the relationship with a simplicity pregnant for implications:

> Originally the state and the temple corporations were the only bodies which possessed the necessary stability and resources for establishing widespread commercial relations. Temple servants were sent on distant missions, provided with letters of credit which enabled them to obtain supplies in other cities. Moreover the temple was the bank of the community through which money could be lent at interest and advances made to the farmer on the security of his crop. Thus in the course of the 3rd millennium there grew up in Mesopotamia a regular money economy based on precious metals as standards of exchange, which stimulated private wealth and enterprise and led to real capitalist development. The temple and the palace remained the centres of the economic life of the community *but by their side and under their shelter there developed a many sided activity which found expression in the guilds of the free craftsmen and the merchants, and the private enterprise of the individual capitalist.* [19]

To put it simply, the fact of the existence of this class is not disputed by reputable historians. And the international extent of this trade should not be overlooked, for trade was known to extend from the city of Ur "over the whole known world

---

18   Astle, *The Babylonian Woe*, p. 39.
19   Christopher Dawson, *The Age of the Gods* (London, 1928), p. 130, emphasis added.

which certainly reached as far afield as Europe, being carried on by means of letters of credit, bills of exchange and 'promises to pay' (cheques), made out in terms of staple necessities; of life expressed *in terms of silver at valuation of barley* (probably at a given season of the year)."[20] This state of affairs clearly implies an international extent to some banking and merchant class, in alliance with the temples of various states. But how would such a class arise and usurp the crown prerogative of money minting, value regulation, and issuance?

2. Ancient Babylon and Egypt

The answer to this question requires a closer scrutiny of the two major military and financial powers of that age: Egypt and Babylon. As noted so far, Babylonian money initially consisted of circulating clay-tablet letters of credit drawn on the surpluses of the state warehouses, and therefore its issuance as money was strictly controlled by the crown. But as also noted immediately above, at some juncture — as early as the third or fourth century B.C., in fact[21] — these existed alongside these clay tablets circulating silver of private issuance, whose value was also fixed in terms of barley or other staples in the state warehouses. In fact, so widespread was the use of the Babylonian system, which also included the use of actual checks, that one finds Babylonians at work in the temples and minting shops of some Greek cities after the Assyrian assault on Babylonia itself.[22] In other words, assaults on Babylon only served to disperse its merchant banking class in other countries, and with that dispersal, the expansion of Babylonian business and banking practices.

a. Early Egypt's Independence from the Babylonian Money Power

It is in this context, then, that the Egyptian money power should be examined, for as has already been noted, Egypt controlled the largest and most productive gold mines — the Bisharee mines — in the ancient world. Here too one notes a parallel development and all its implications:

> According to Breasted, gold and copper rings of a fixed weight circulated in large scale business in the time of the "Old Kingdom," and (significantly enough to the student of "banking," or private money creation and regulation, as it might better be known) "stone weights were already marked with their equivalence in such rings." The

---

20 Astle, *The Babylonian Woe*, pp. 13–14, emphasis added, citing Sir Charles Woolley, *Abraham* (London, 1936), pp. 124–125.
21 Astle, op. cit., p. 12.
22 Ibid.,, p. 55.

circulation as money of these "promises to pay" recorded on stone, pointedly suggests the likelihood of the activities of a secret fraternity whose hereditary trade was private money creation.[23]

Indeed, the parallel with Babylon is almost identical, since the latter used clay tablets as units of exchange, while Egypt used stone weights marked with their equivalent value in precious bullion. But there is likewise a significant difference as well, for in Babylon's case those clay tablets circulated as money prior to the issuance of bullion-based monies, and the value of such monies was stated in terms of stores in the state warehouse. But with Egypt, one finds a later stage of development, for the value of the stones is expressed, *not in terms of the surplus stores in a warehouse, but in terms of bullion itself.* In short, the concept of money has been *degraded,* and with it, there is almost no visible connection to the state and crown prerogative of money regulation and issuance, not to mention anything of practical value as the result of industry or creativity, i.e., goods.

And this degradation of money in Egypt, i.e., the fact that its promises to pay are expressed in valuation of bullion, rather than bullion being expressed in valuation of real goods, plus the fact of the existence of Egypt's Bisharee mines with their international contingent of mercenary guards, points to a very significant and — because of its subtlety — often overlooked point:

> As a result, although the Egyptian empire in the earliest years *might very well be described as a common market existing independent of the Babylonian money power, and deriving its strength from the will to be of a dedicated and instructed Ruler,* the sequence of events shows that through those concessions it obtained for its best services in war, it was *not long before international money power re-penetrated the substructure of Egyptian life and established its usual behind-the-scenes influence,* if not control, as in the earlier time that denoted the collapse of the "Old Kingdom." It may safely be considered to have reassumed the position of hidden power it had held a thousand years before during the closing years of the 6th dynasty, a period in which the stone weights indicating equivalence in metal money circulated in much the same way as clay facsimiles of contemporary coinages circulated...[24]

To put it differently, Egypt, once under the influence of that international bullion brokers' power, had managed to throw off that influence, only to have it restored nearly a millennium later.

---

23 Astle, *The Babylonian Woe,* p. 36, citing James H. Breasted, *A History of Egypt,* pp. 97–98.
24 Astle, *The Babylonian Woe,* p. 43.

3. The Conspirators at Work
a. Economic and Military Autarchy and Modern Analogues: Sparta and the Greek Tyrants

But how, exactly, does one *know* that such an international money power existed in ancient times? Astle's answer is as surprising as it is both contemporary and simple: one had only to look at certain ancient wars in conjunction with a look at the economic activity and policies of various states to discern a hidden hand at work. And the best example of this war between a money power and a state determined to maintain its independence from it, its *autarchy*, are the Peloponnesian Wars between Athens and its allies, and Sparta. It is worth citing Astle and his sources at length:

> Of the tyrants of Greece and Asia Minor in ancient times, the learned Professor Heichelheim wrote:
> "These tyrants were for the most part members of the nobility themselves who had made the grade using the new political and economic possibilities of their time to overthrow their own equals and to subdue their whole state temporarily. *The tyrants were often compelled to introduce the coin economy pattern into the area over which they ruled, or at least to promote its development officially, in order to gain the upper hand over their enemies...*

(One wonders: Compelled by *whom?*)

> ...to stabilize the position of the peasantry on the land, and to expand and rebuild state economy, *a central distribution of money and goods in kind partly directed towards mercenaries,* bodyguards and various political friends and partly indirectly to the masses of poor people in the form of wages paid for extensive building operations and improvements, is characteristic of tyrant economy...[25]

Astle comments as follows:

> The above remarks of Professor Heichelheim indicate there were "new political and economic possibilities: in that period 650–500 B.C. when the tyrannies most of all flourished… The question then becomes, what were these "new political and economic possibilities"?... The answer is arrived at readily: they derived from the activities of

---

25 Citing *Encyclopedia of World History* (Houghton Mifflin, publishers, Boston, 1940), p. 48, cited in Astle, *The Babylonian Woe,* p. 96, emphasis added by me.

the agents of the international silver bullion brokers who, from ports such as Argos, Athens, and Aegina where King Pheidon struck the first Greek silver coinage ca. 680 B.C., promoted the luxury traders who sold their wares from wigs to harlots as against the new silver coinage or promise thereof. The opportunities *clearly were for those who assisted in the monetization of the city, and all its activities and possessions, and its population....*[26]

The result of this "monetization" of the city-state — and of the reduction of each member of its population to the status of a mere "human resource" (to use the modern term) — was, according to Professor Heichelheim, entirely predictable, and eerily reminiscent of the "free trade" agreements of the late 1980s and 1990s and of their disastrous effects on the industrialized societies that adopted them (and remember, Heichelheim was writing in the 1940s, long before such agreements were in the offing, and was speaking of *ancient times!*):

The aristocracies refused equality to the landless traders and manufacturers, the peasants were oppressed by the rich and encouraged to get into debt and then were reduced to slavery and exile; slaves began to compete with free labour. Ambitious individuals capitalized this discontent to overthrow the constituted government and establish themselves as tyrants in all the Greek cities *with the notable exception of Sparta.* [27]

So what is so significant about Sparta that the rest of the Greek city-states went to war with her?

Astle puts it "country simple":

...(T)he state that rejected international money power, as did Sparta and Rome in ancient times, and Russia in modern times, had to be prepared to establish total military self-sufficiency.[28]

One might add Nazi Germany to Astle's list, for like Russia and ancient Sparta, the regime of the Third Reich had determined not only to restore the issuance of money to the State, but had embarked on a program of military and energy autarchy as well.[29]

---

26 Astle, op. cit., p. 96, emphasis added.
27 Citing *Encyclopedia of World History* (Houghton Mifflin, publishers, Boston, 1940), p. 48, cited in Astle, *The Babylonian Woe*, p. 96, emphasis added by me.
28 Astle, *The Babylonian Woe*, p. 40.
29 For the aspect of energy autarchy and the allied military technologies, see my *The SS Brotherhood of the Bell, Secrets of the Unified Field*, and *The Nazi International*.

So what, exactly, had Sparta done that so infuriated the rest of the Greek city-states? The answer is complex, and requires a brief detour. Both Plato[30] and Aristotle[31] were clear that money had value only insofar as it was defined by the law of the state; in other words, they both obliquely recognized that the power of issuing and regulating the value of money was a state, not a private, prerogative, and this in spite of the fact that by their time, the private issuance of money was well-established practice.[32] Thus, when one turns to the city-states of the tyrants, and to their silver-coin issuances, one discerns a clear, though carefully disguised, hidden hand:

> The evidence that the earliest coinages in Greece had essentially a local circulation in no way alters the picture previously outlined of silver money as being part of an international conspiracy. All Greek states apart from Athens and Samos, Siphnos and Corcyra, and possible one or two others, had to obtain silver bullion for their coinage from abroad, which necessarily obliged them to deal with those traders who specialized in dealing in bullion. Such trade in bullion had to be in the hands of a small and highly secretive group, as much on account of the sources of supply being relatively few and scattered as it were out to the ends of the earth, as on account of the fact that it would be only such a group that could also control those supplies of slave labour and their purchase from triumphant peoples whose warlike activities, as likely as not, they had instigated themselves; slave labour so necessary to the success of their mining operations.[33]

Once the course had been determined in various Greek city-states to issue bullion-based money, which bullion they had in scarce supply, those city-states came increasingly under the influence and sway of the bullion brokers themselves. All except one, which refused to issue bullion-based coinage.

Sparta.

Hence one has now a basis of understanding for Sparta's well-known harshness and the military discipline it imposed on her whole population, for if she were to maintain her independence of the money power infecting the other Greek city-states, her military had to be the best. Thus, there is a hidden, and much more real, purpose to the infamous Peloponnesian War — a war that neither Athens nor Sparta "had won and neither one had lost." The war left both powerful Greek city-states "exhausted, and over their

---

30   Astle, op. cit., p. 139.
31   Ibid., pp. 139–140.
32   Ibid., p. 140.
33   Astle, *The Babylonian Woe,* p. 140.

prostrate bodies the servants of this same sardonic Money power drew the chains of their slavery."[34] That real underlying purpose of the Peloponnesian War "was to establish private common money market across the Greek world totally controlled by the *trapezitae* or bankers in modern terminology."[35] Consequently, "judging by Sparta…and later by Rome itself, it seems that in ancient times there was some considerable understanding of the power inherent in precious metal money to destroy, by lending itself to manipulation, the *status quo* of any race or state."[36]

b. Babylon, Persia, and Money Creation

Yet another, and even more persuasive, example of the actions and hidden directions of a money power on the international scene is afforded by the case of Babylon and Persia. It is worth citing Astle at length once again:

> It is not until the Assyrian, Neo-Babylonian, and Persian eras that clear evidence can be traced of the total degeneration of kingly power and of kings and so-called emperors as quite often being little more than gloriously be-medalled front men for private money creative power striving to create world-wide hegemony… They still continued to be needed principally as a point towards which the eyes of the people might be diverted in order that the people might not realize that all was not well in that direction towards which their loyalties naturally leaned, nor glimpse the destructive forces that were gnawing at the roots of the Tree of Life itself. Even as far back as 2500 B.C. Sargon of Akkad proceeded into Anatolia to chastise the city of Ganes on account of the commercial community of Mesopotamia; probably to enforce the payment of interest on loans, or repayment of principal… One of the reasons of the success of Cyrus, though but a petty Persian prince formerly to 550 B.C. when he deposed his sovereign, Astyages the Mede, is clear from the circumstances of his victory over Croesus of Lydia in 546 B.C.
>
> Croesus had offended the international money powers by seizure of their treasure held by their agent Sadyattes, and by the total assumption of monetary issue by the state. Example had to be made of him to deter other princes from similar action, and the eager and ambitious Cyrus was obviously the one chosen for this purpose. According to the article on Babylonia in the *Encyclopedia Britannica*, 9th edition, by Professor Sayce, Croesus had rashly joined battle with Cyrus without waiting for

---

34 Ibid., p. 143.
35 Ibid.
36 Astle, *The Babylonian Woe*, p. 144.

the arrival of his Babylonian allies under Nabu-Nahua the father of Belshazzar of the Book of Daniel. It is more than likely however, that a truer reading of these events would be that international money power, patron of the rise of Cyrus both through organization of his supplies of mercenary soldiers, and of the best of weapons, had been the principal influence in these events as in other enterprises of Cyrus, such as the siege of Babylon 14 years later. Thanks to his influence, while the progress of Nabu-Nahua towards junction with the forces of Croesus would have been sabotaged, Croesus himself would have been misinformed of the intentions and strength of both Cyrus and Nabu-Nahua.

Cyrus won the day, and Croesus was totally humbled. Having thus proven his "suitability," and his readiness to promote the policies of his financial backers, the relatively easy conquest of Babylon was arranged for Cyrus some fourteen years later.[37]

Who arranged for the easy conquest of Babylon? More importantly, who financed the vast Persian war machine?

The answer, according to Astle, is a biblical one. "It is interesting to note," he observes,

> That shortly after the entry of the Persian forces into the city, the "Children of Israel" were permitted to return to that which they considered their homeland, and every assistance was given them towards renewal of their national life and the rebuilding of their temple, which, of course, was its heart.... The special concessions made by Cyrus to the Hebrew almost on entry into the city of Babylon, would certainly suggest that he had received their substantial assistance, perhaps through financing towards the purchase of the finest of military accoutrements such as would be obtainable only through the good graces of the Babylonian commercial and banking houses, or through that information with which the Hebrew may have kept him constantly supplied such as the state of military preparedness within the city, etc.
>
> It may reasonably be assumed that the Babylonian money power was completely international in outlook, whatever its outward profession, and totally unsympathetic towards the ancient faith of the Ziggurat and the worship of Marduk, and towards the intended effects of the restoration of the Ziggurat at Ur, at that time, by Nebuchadnezzar. If in earlier Assyrian times such money power certainly was not the Hebrew, though possibly linked thereto through members of the

---

37 Astle, *The Babylonian Woe,* p. 71.

latter Israelitish Confederacy... the fact of the existence of powerful Hebrew influence in international finance in Neo-Babylonian times seems a reasonable supposition.

The Hebrew...may have risen to especially privileged position in the Babylonian money industry, if that is what it can be called, and may have come to learn at that time those secret practices of the money changers' craft, which he was certainly forbidden in his native land, according to the Laws (of) Moses.[38]

Indeed, as Astle points out, the Hebrews returning to Palestine were compelled, under Ezra's edict, to divorce their foreign wives. While certainly explainable as a religious ordinance designed to restore doctrinal and racial purity, Astle notes that the practice of intermarriage is itself the hallmark of the international banking class to this day, and thus, Ezra's ordinance may itself be an indicator that the Hebrews had indeed penetrated the international money power during the Babylonian captivity.[39] There is yet another indicator of this "babylonization" of Hebrew religion and practice, and that is the existence of the Talmud and the rabbinate itself, and the adopted symbol of the Hebrews and Jewish people to this day, the so-called *Mogen David* or Star of David: ✡. This symbol is itself of Babylonian, not Hebrew, origin, and as we shall see in the next chapter, it is an apt symbol of a fraternity that is "keeping secrets."

C. How the Conspiracy Worked

It is perhaps in the actual history of the development of the private issuance of bullion-based monies in ancient times and in various civilizations that one most clearly discerns the evident hand of a hidden international class of bullion-brokers, war merchants, slave traders, and mining operators, for almost invariably, the pattern is the same.

1. The First Stage: Penetrate and Ally with the Temple

As noted repeatedly throughout the previous pages, from Greece to Rome to Egypt and the Mesopotamian civilizations of Sumer, Babylon, and Assyria, there is a clear association of banking activities, and of minting itself, with the temple. The reason for this association, at least as far as banking and finance go, are easy to rationalize:

These persons had conducted their business in the shade of the temple courtyards from ancient days as, and if they could, in order that the

---

38 Astle, *The Babylonian Woe*, p. 76. See also p. 129.
39 Ibid., pp. 76–77.

power or mystery as locally was held in awe, might give sanctity to their activities which so often were exercised against the well being of the people who sheltered them....

These agents would have lurked as only faintly discernible shadows behind the temple façade, although they instigated much of what came to pass in those days, if themselves so little seen. Of first concern to them would have been the reputation of their masters, the priesthood, for piety, probity, and godliness, in so far as appearance went. For by maintaining the position of the priesthood, they maintained themselves and their secret power; yet for whatever they brought about, especially if of evil, it may safely be assumed, a nevertheless inviolate priesthood would be held responsible.

With the sanctity of the temple as a shelter and cloak of their activities, this bullion brokers' class could then engage in the next stage of their activity:

2. The Second Stage: Issue False Receipts

That second stage is to issue counterfeit or false receipts against the credit or surpluses of the state warehouse, and *thereby to increase the money supply itself:* the principal device at this stage "may be assumed to be the secret and private expansion of the total money supply effected primarily by the issuance into circulation of false receipts for...valuables supposedly being held on deposit in thief-proof vaults, or otherwise, for safe custody."[40] Such issuance would be relatively easy to come by, for by penetration of and close association with the temples of various states, they would have had access to the seals and other devices used to authenticate letters of credit by the temple priesthoods.

3. The Third Stage: Substitute Bullion for Letters of Credit as a Measure Against False Receipts

This brings us inevitably to the third stage, for once a sufficient number of such false receipts were in circulation, it would become apparent both to the crown and to members of the temple itself that there were simply too many false receipts in circulation, and that a monetary reform was necessary. At this juncture, in step the bullion brokers with the solution: holding a virtual monopoly on mining, they propose the substitution of bullion-based coinage, whose value is initially defined by a certain amount of actual goods — barley, as we saw in the case of Mesopotamia — for the actual state-issued letters of

---

40 Astle, *The Babylonian Woe*, p. 13.

credit. Once any crown and temple alliance within any given state accepts this arrangement, it will become increasingly hostage to the need for bullion and for the merchants, slave traders, and mining-mercenaries that can supply it, and that have the technology to mine and melt and mint it.[41]

4. *The Fourth Stage: Then Create a Facsimile of Money*

Having artificially stimulated and increased the value of bullion, this class then creates the final step: it issues new letters of credit *against the supply of bullion rather than against the warehouse surplus,* and defines those letters of credit as valuation in terms of weight of some precious bullion. Thus, we arrive back at the state of issuing false receipts, only in this instance, these receipts are neither false nor counterfeit, they are simply privately created notes or promises to pay a certain amount of bullion (which they also control). Thus, once again the money supply is not only expanded, but the use of such instruments actually served to allow the bullion brokers to circulate more of such notes than they had actual bullion to redeem. The key, once again, was the sanctifying probity that the temple association gave them. With this step, their power and influence over the various states which they penetrated was almost complete, for it gave these ancient bankers, like their modern counterparts, the ability to expand or contract a state's money supply, and to control their economies to create boom or bust. The final step, in other words, was to create a clay or stone (or in modern times, paper or plastic) *facsimile of the facsimile of state-issued money* (the bullion-based monies, also of their own creation)!

D. THE PARALLEL TO THE AFTERMATHS OF THE COSMIC WAR, AND WORLD WAR II: THE GLOBALIST AGENDA AND MONEY, MONARCHIES, MONOTHEISM, AND MILITARIES

But there is a final piece to this ancient story and plot, and it is one of the most breathtaking of them all. The effect of the alliance of this bullion brokers' class with the temple throughout the ancient world would have brought to their attention the fact that the pantheons of various civilizations, and even the stories and acts associated with various gods, were all remarkably similar. As a result of this realization, the bullion brokers would also come

> To realize that they could actually create that which functioned as money with but the record incised by the stylus on the clay tablet promising metal or money. Obviously, as a result of this discovery

---

41 See Astle's discussion on pp. 15ff, *The Babylonian Woe*.

which depended on the confidence they were able to create in the minds of the peoples of their integrity, provided they banded themselves together with an absolute secrecy that excluded all other than their proven brethren, they could replace the god of the city himself as the giver of all. *If so be they could institute a conception of a one god, **their** god, a special god of the world, a god above all gods, then not merely the city, be it Ur or Kish or Lagash or Uruk, but the world itself could be theirs, and all that in it was.*[42]

In other words, behind the creation of the vast empires of Persia, and later the Macedonian-Greek empire of Alexander the Great, there lies the activity of a powerful financial interest, striving to create an ancient "New World Order" with a central state and unified religio-cultural world.[43] And the mechanism by which to foster this religious worldview would be to gradually point out, and inculcate the idea in popular imagination, that all local religions were but the same thing, various "names" for the "same gods."[44] Thus, the whole weight of their combine and cabal would have been thrown behind the promotion of a monarchial front, representative of the new empire and the new monotheism, and backed by the latest in military hardware.

But there is another possibility, one suggested by the scenario of a cosmic war outlined in the first chapter,[45] and that is, if the survivors of a once scientifically sophisticated civilization were ever to reach the pinnacle of scientific power and achievement that would have made an interplanetary war *possible,* if, that is, they were ever to reconstruct the destructive technologies of their hegemony and extend themselves again into space, then they would have quite literally to draw on the full resources of the entire world, and to create the wars and conflicts that drove technological achievement forward at a faster-than-normal pace of development. In this case, they would develop not only secret associations for this purpose, but money, and a close association with the temples, would be the easiest and least technologically sophisticated method to do it. And the political goal would be the same: ever-larger empires, eventually to encompass the entire globe itself.

This association of money creation with the temples, and with the goal to create a unified world order, for the purpose of reconstructing a lost technology of hegemony, is where the story gets *really* interesting. As they say, the plot is about to thicken....

---

42   Astle, *The Babylonian Woe,* p. 15, italicized emphasis added, boldface emphasis in the original.
43   See also Astle, op. cit., p. 192.
44   In this regard, one need only consider the modern parallel, with Rockefeller financing of liberal Protestantism and various ecumenical agencies and schemes.
45   The whole Cosmic War scenario I have outlined in *The Cosmic War: Interplanetary Warfare, Modern Physics, and Ancient Texts* (Adventures Unlimited Press, 2007).

Six

ALCHEMY UPSETS THE APPLECART
THE TRANSMUTATIVE MEDIUM AND THE ALCHEMY OF THE STARS AND BANKSTERS

❖

> *"Further support for the hypothesis that a transnational interrelationship existed between ancient Mystery Schools, was the discovery of the Gunderstrup Cauldron — a superb example of the silversmith's art. This bowl, which was discovered in a Jutland peat bog, is decorated with pan-cultural deities."*
> —Brian Desborough[1]

So once again, what *possible* connection could there be between all these things? Why *do* we find a J.P. Morgan suppressing Tesla? Why in turn do we find a Tesla preoccupied with "gripping the earth" to make his "wireless transmission of power" work? Why *do* we find a startling ancient connection between banking and physics (in the form of astrology, astronomy, and even — as Hoagland has hinted — sacred geometry)? Why *do* we find an RCA engineer named Nelson drawing up planetary charts that look for all the world like astrological charts? Why do we encounter a government economist like Edward Dewey devoting a lifetime of work to the study of cycles of all sorts? And why do we find him making conscious and deliberate reference between economic activity and physics? And why do we find, not just in ancient times, but in modern ones as well, the presence of bankers on the peripheries of such investigations?

---

1 Brian Desborough, *They Cast No Shadows: A Collection of Essays on the Illuminati, Revisionist History, and Suppressed Technologies* (San Jose, California: Writers Club Press, 2002), p. 175.

A. Econophysics
1. Physicists Invade Finance: The Modern Model as a Key to the Paleoancient Past

The answer lies in part in modern times, and in order to begin to peel back the onion skin layers of this vast, intricate, and ancient relationship, one must look at the rise of a new discipline — "econophysics" — and at what happened in modern times as physicists entered the world of high finance, bringing their techniques and tools of applied mathematics and analysis with them.

The term "econophysics" was actually coined in the mid-1990s by H. Eugene Stanley[2] to reflect two things: first, it was designed to reflect the vast influx of degreed physicists into the financial sector itself, as that sector often provided a higher salary base and opportunities to employ the tools of applied mathematical analysis than was possible in actual standard academic or experimental careers, and secondly, it was reflective of the fascination that physicists had acquired for the application of the techniques of statistical analysis they had developed in quantum mechanics to the problems of economic and financial analysis itself.

As such, econophysics "is an interdisciplinary research field, applying theories and methods originally developed by physicists in order to solve problems in economics, usually those including uncertainty or stochastic processes and nonlinear dynamics."[3] What spurred this interest was in part, according to the article in Wikipedia, the "availability of huge amounts of financial data, starting in the 1980s."[4] Of course, as has been noted, this is not exactly true, since there was a vast amount of data available from the Department of Commerce and from other sources and countries dating much further back, a database utilized by Edward and the Foundation for the Study of Cycles, and a database known to its members, some of whom included economists from major international banks.

One thing in particular that attracted physicists to this study of the application of their techniques to financial and economic systems was precisely that in economics, more often than not, there prevailed a general condition of non-equilibrium,[5] that is, one was never dealing with systems where there was a uniform distribution of resources, wealth, and so on. As physicists and chemists had only recently — in the 1960s — begun to study non-equilibrium physical systems and their remarkable ability to self-organize, the motivation for their study of economics becomes clear: physicists thought that by

---

2 "Econophysics," en.wikipedia.org/wiki/Econophysics, p. 1.
3 Ibid.
4 Ibid.
5 Ibid.

studying the most notoriously non-equilibrium systems known to mankind — economic systems — their models of the ability of such systems in the physical world could be applied with some success to model self-organization in economic systems, and to make valuable predictions on that basis. As will be seen in a few moments, this is an important clue to the possible deeper physics involved in economic systems.

To put it succinctly, quantum mechanics had invaded economics, and the reason why is relatively easy to perceive, because since economic activity was "the result of the interaction among many heterogeneous agents, there is an analogy with statistical mechanics, where many particles interact."[6] And the key tool in this new interdisciplinary endeavor was quantum physics' use of the "path integral formulation of statistical mechanics."[7]

But what exactly does this mean?

The first implication is that one may use the influx of physicists into economic study in modern times as an analogy of what happened in ancient times, and assume that something like this happened as well in ancient times, as those with the "high knowledge" of a deeper physics penetrated the world of high finance, that is to say, the temple, seeking a better standard of living and personal wealth, like their modern descendants. And once inside those hallowed pavilions and chambers, they allied with the bankers of their day, the "bullion brokers." It was a necessary détente in the aftermath of that great Cosmic War, for if the lost science and technology were ever to be reconstructed, it would require lots of money, and lots of scientific expertise, and both resided in ancient times in the temple.

The second implication requires a closer look at quantum mechanics itself.

2. Quantum Mechanics, Ancient Astrology, and the Statistical Approach

One may gain an appreciation of the statistical nature of quantum mechanics by considering its foundational principle, the Heisenberg Uncertainty Principle, itself. Briefly stated, the Uncertainty Principle holds that if one measures the velocity of an electron, one cannot measure its position, and conversely, if one measures its position, it is impossible to measure its velocity. Thus, quantum mechanics, when considering the behavior of *several* particles all at once, had to rely upon observation and the compilation of statistical probabilities in order to model that behavior.

The same holds true, if one takes the ancient texts at face value, for astrology as well, for there is an exact modern analogy with quantum mechan-

---

6 Ibid., p. 2.
7 "Econophysics," en.wikipedia.org/wiki/Econophysics, p. 2.

ics and its countless and untold observations made diligently over time, and with its statistical approach to that vast number of interactions, for this is precisely what one encounters in ancient times as well. In fact, as previously noted, the Babylonian "omen tablets" indicate that astronomical observations of planetary alignments and corresponding earthly activities and effects — a combination we now call astrology — were made over hundreds of thousands of years ago. This is therefore yet another sign that there is some other hidden influence at work in the temples besides that of the bullion brokers, and that influence is that of the ancient astrologers-astronomers themselves, compiling their lists of observations. From the previous chapters it is evident that the ancient international "bullion brokers" were in league with the various temple priesthoods-astrologers.

3. Dr. Li's Gaussian Copula and a Physics Analogue: The Multibody Problem

Quantum mechanics' statistical approach, and that implied by the millennia of ancient astronomical and astrological observation, is directly connected to Dr. David Li's Gaussian copula formula via a rather interesting route: the multibody problem. The problem is an exact analogue to the difficulty of modeling the behavior of mass aggregates of particles. The only way to do so was through careful and countless observations and a statistical model of probabilities of their behavior. But the problem becomes more acute as the size of bodies being modeled grows.

Imagine one is trying to model not the aggregate behavior of several particles, but of several planets and stars. It is relatively easy to model the force of gravitational attraction between two moving masses or planets, and to predict their behavior. But as one adds more and more moving masses to the system to be modeled, the ability of mathematics to handle and accurately predict the behavior of a total dynamic system or any of the individual components within it progressively breaks down, and margins of error increase. Dr. Li, of course, solved his "multibody problem" of correlations of economic activity by, predictably enough, a statistical approach, which introduced the problem of random probabilities into economic activity.

But what if the randomness of quantum mechanics was itself the result of a deeper physics? And, if that be so, and since "econophysics" was but the importation into economics of physical models, would that deeper physics also be a deeper physics of economic activity?

4. The Deeper Physics:
a. David Bohm's Hidden Variable Quantum Mechanics and The Implicate Order

The deeper physics implied by this question may be readily perceived by a review of one of its most famous expositors, the well-known plasma and quantum physicist David Bohm. Bohm is best known, in fact, for his "hidden variable" version of quantum mechanical theory, in which the "randomness" of quantum mechanics observed and measured by physicists (Bohm's "explicate order") is understood to point to a deeper and more ordered hyper-dimensional reality (Bohm's "implicate order"). Bohm outlined these views in a popularized treatment in a book entitled *Wholeness and the Implicate Order*, and it is this book that will be the basis of our review of his ideas.

For Bohm, the whole development of quantum mechanics pointed to the real existence of this hyper-dimensional "implicate order":

> One discovers...both from consideration of the meaning of the mathematical equations and from the results of the actual experiments, that the various particles have to be taken literally as projections of a higher-dimensional reality which cannot be accounted for in terms of any force of interaction between them[8]

In other words, Bohm is saying almost exactly the same thing as was Richard C. Hoagland at the beginning of this chapter: particles, as rotating masses, and whose behavior when measured as a statistical aggregate, are *portals* that "gate" a hyper-dimensional reality into our own world. While Hoagland, however, was concerned with the hyper-dimensional implications of the physics of very large masses, i.e., stars, Bohm is concerned with the hyper-dimensional implications of the physics of the very small. This is an indicator that the physics involved is scale-invariant, that is, applicable over the whole range of scales of objects with which physics deals.

Bohm's apt analogy of the interaction of this hyper-dimensional world with our own is that of a projection. "Let us begin," he says, "with a rectangular tank full of water, with transparent walls."[9] He then reproduces a diagram of a fish tank, with two television cameras, one pointed at one side of the tank, and the other at the side perpendicular to each. Each camera, "A" and "B" respectively, is attached to one television monitor. As the fish swims around, the images from the two cameras are sent to the monitors. Bohm then comments as follows:

---
8 David Bohm, *Wholeness and the Implicate Order* (London: Routledge, 1999), pp. 186f.
9 David Bohm, *Wholeness and the Implicate Order*, p. 187.

> What we will see there is a certain *relationship* between the images appearing on the two screens. For example, on screen A we may see an image of a fish, and on screen B we will see another such image. At any given moment each image will generally *look* different from the other. Nevertheless the differences will be related, in the sense that when one image is seen to execute certain movements, the other will be seen to execute corresponding movements. Moreover, content that is mainly on one screen will pass into the other, and vice versa (e.g., when a fish initially facing camera A turns through a right angle, the image that was on A is now to be found on B). Thus at all times the image content on the other screen will correlate with and reflect that of the other.[10]

In other words, our three-dimensional world acts as a sort of "prism" to break up a singular hyper-dimensional object into a fragmented thing, perceived from different angles but with correlated movements.

Bohm summarizes this view of our three-dimensional reality as a projection from a high-dimensional reality, and its effect on perceptions of quantum mechanics and its underlying implicate order, with words eerily evocative of Hoagland's comments about stars at the beginning of this chapter:

> …(W)e may regard each of the "particles" constituting a system as a projection of a "higher-dimensional" reality, rather than as a separate particle, existing together with all the others in a common three-dimensional space. For example, in the experiment of Einstein, Podolsky and Rosen, which we have mentioned earlier, each of two atoms that initially combine to form a single molecule are to be regarded as three-dimensional projections of a six-dimensional reality.[11]

However, it is when Bohm turns to a consideration of the implications of this view that the full scope of the possibilities of this deeper physics springs into full view, and with them, the reason for the association of banksters with temples in ancient times — not to mention modern times! — becomes a little clearer.

To make these implications manifest, Bohm begins by noting that

> …(W)hen the quantum theory is applied to fields… it is found that the possible states of energy of this field are discrete (or quantized). Such a state of the field is, in some respects, a wavelike excitation spreading out over a broad region of space. Nevertheless, it also has somehow a discrete quantum of energy (and momentum) propor-

---

10 Ibid.
11 Bohm, *Wholeness and the Implicate Order,* p. 188.

tional to its frequency, so that, in other respects it is like a particle (e.g. a photon). However, if one considers the electromagnetic field in empty space, for example, one finds from the quantum theory that each such "wave-particle" mode of excitation of the field has what is called a "zero-point" energy, below which it cannot go, even when its energy falls to the minimum that is possible. If one were to add up the energies of all the "wave-particle" modes of excitation in any region of space, the result would be infinite, because an infinite number of wavelengths is present.[12]

Before we can proceed, it is necessary to pause and consider how Bohm has characterized the zero-point energy. For Bohm, as the passage just quoted makes clear, this energy is a result of the quantized nature of space itself, and of the fact that this quantization is the result of wavelike structures — of areas of compression and rarefaction — within it.

Why this is so requires only a moment's reflection. When a physicist suggests that space itself is quantized, what he means is simply that it is not an infinitely divisible continuum, capable of being divided into ever smaller units or cells *ad infinitum*. The reason is, according to Bohm, that if space itself is the result of wavelike structures of compression and rarefaction, then such wavelike structures will inevitably induce cells within it. What appears to be an infinitely divisible continuum, therefore, is really so only because of the infinite number of wavelengths present at any given point of vacuum space.

This conception leads Bohm to posit the next step toward the quantization of space itself:

> ...(T)here is good reason to suppose that one need not keep on adding the energies corresponding to shorter and shorter wavelengths. *There may be a certain shortest possible wavelength,* so that the total number of modes of excitation, and therefore the energy, would be finite.[13]

We will refer to this "shortest possible wavelength" as the "Bohm wavelength."

This "Bohm wavelength" — the idea that wavelike phenomena are themselves quantized — has profound and huge implications for the type of "deeper physics" being explored here. For one thing, this unknown wavelength sounds very much like the "lost word" of Masonic tradition, or the "lost chord" of the esoteric doctrine of the harmony of the spheres, the lost chord being the frequency which somehow binds all of physical creation together, *since the latter is but a harmonic or overtone series of it.*

---

12 Ibid., p. 190.
13 Bohm, *Wholeness and the Implicate Order,* p. 190, emphasis added.

This conception turns out to have huge implications for the possible unification of the physics of the very large with the physics of the very small:

> Indeed, if one applies the rules of quantum theory to the currently accepted general theory of relativity, one finds that the gravitational field is also constituted of such "wave-particle" modes, each having a minimum "zero-point" energy. As a result the gravitational field, and therefore the definition of what is to be meant by distance, cease to be completely defined. As we keep on adding excitations corresponding to shorter and shorter wavelengths to the gravitational field, *we come to a certain length at which the measurement of space and time becomes totally undefinable [sic]*. Beyond this, the whole notion of space and time as we know it would face out, into something that is at present unspecifiable. *So it would be reasonable to suppose, at least provisionally, that this is the shortest wavelength that should be considered as contributing to the "zero-point" energy of space.*[14]

In short, *find that wavelength and one will find the ability to engineer space-time and all that is in it.* One would be able to tap into a virtually inexhaustible source of energy and utilize it for whatever purpose one desired. Additionally, Bohm's remarks also suggest that finding this frequency would also be a step on the road to the manipulation of gravity itself.

Bohm, moreover, clearly sees this implication of engineerability, for he sees the clear implication of his view for the very "construction and constitution" of matter itself. Indeed, he even goes so far as to give an approximation of the very frequency that constitutes the "Bohm wavelength"!

> When this length is estimated it turns out to be about 10^{-33} cm. This is much shorter than anything thus far probed in physical experiments (which have got down to about 10^{-17} cm or so). If one computes the amount of energy that would be in one cubic centimetre of space, with this shortest possible wavelength, it turns out to be very far beyond the total energy of all the matter in the known universe.
>
> What is implied by this proposal is that what we call empty space contains an immense background of energy, *and that matter as we know it is a small, "quantized" wavelike excitation on top of this background, rather like a tiny ripple on a vast sea.*[15]

---

14 Bohm, *Wholeness and the Implicate Order,* p. 190, emphasis added.
15 Ibid., pp. 190–191, emphasis added.

Before pondering the implications of all this further, it is worth citing Bohm on the awesome power implied in such a view:

> It is being suggested here, then, that what we perceive through the senses as empty space is actually the plenum, which is the ground for the existence of everything, including ourselves. The things that appear to our senses are *derivative forms* and their true meaning can be seen only when we consider the plenum, in which they are generated and sustained, and into which they must ultimately vanish.
> ...
> In our approach (the) "big bang" is to be regarded as actually just a "little ripple." An interesting image is obtained by considering that in the middle of the actual ocean (i.e., on the surface of the Earth) *myriads of small waves occasionally come together fortuitously with such phase relationships that they end up in a certain small region of space, suddenly to produce a very high wave which just appears as if from nowhere and out of nothing. Perhaps something like this could happen in the immense ocean of cosmic energy, creating a sudden wave pulse, from which our "universe" would be born.*[16]

Observe quite carefully what Bohm has just stated, for it is crucial to all that follows: *matter itself, in all its variegated forms, is the result of an interferometry — of the "mixing" — of several waves of various wavelengths, all of them in turn harmonics or overtones of the "Bohm wavelength." It is the very technology of the creation ex nihilo, of the ability of the physical medium to create information and systems by wave-mixing. It is the technology of creation from nothing: alchemy.* And there is one final consequence of Bohm's views, and that is that in order to give relative system stability to such waves, or rather to the systems that such waves generate, the easiest way to do so is via rotation.

A consequence inevitably follows from this view, and that is that rotating material systems — whether stars or particles — are thus *natural resonators of such waves*, and hence to understand the pattern of interference of such waves, one had therefore *to monitor the geometrical positions of significant objects in local space-time.* In short, Bohm, with his view that matter is a "portal" and a glimpse into this higher-dimensional reality and its energies, has provided a rationale for two things: the ancient preoccupation with *astrology, and the ancient preoccupation with alchemy, for both are manifestations of one and the same physics.* Matter, to put it succinctly, is but a standing wave of interferometry of other waves; matter itself is a *grid or template of interference* of such waves. This forms the link to

---

16 Bohm, *Wholeness and the Implicate Order,* pp. 191–192, emphasis added.

alchemy, for as such, matter emerges as information within the field of the physical medium. To put it succinctly, matter is mutable, or in the language of alchemy, *transmutable*, since the physical medium itself, the Philosophers' Stone par excellence, transmutes itself into the diversity of the material creation. This idea of matter as *a template of the interferometry of such waves* will play a key role when we turn to examine the placement of sacred temples along a certain earth grid in the next chapter, and an even more crucial role as a connecting concept typing together all the disparate data points in this book upon its conclusion.

b. The Foundation for the Study of Cycles Notices a Similar Thing

Oddly enough, the Foundation for the Study of Cycles noticed a very similar thing. We have already encountered the fact that the Foundation's founder, Edward Dewey, compared the many cycles in its database to wave forms, and noticed the fact that various waves of cycles could be "averaged" together, like sound waves.

Ray Tomes, a member of the Foundation, wrote an interesting paper presented at the Foundation's February 1990 conference. The paper's title is pregnant with implications: "Towards a Unified Theory of Cycles." In it, Tomes pursues Dewey's and Dakin's sound wave analogy to a breathtaking conclusion:

> Eventually I realised that the pattern of frequencies present in ... corn prices *was the same as the arrangement of frequencies of the white notes on a piano...* This was peculiar, and going back to my early common economic cycles study I realised that the ratios 4:5:6:8 were exactly those of a major chord in music! Why are economic series playing major chords and scales in **very** slow motion?
>
> Research showed that such patterns had been observed and reported before by several contributors to *Cycles* magazine.[17] One of these was D.S. Castle (1956) who found that stock market cycles fit the musical scale. The pattern found ranged over three octaves, and all seven white notes plus one black note were present in at least one octave.[18]

Tomes' suggestion, in other words, is simply one and the same as that of David Bohm, namely, that one might find "the ultimate wavelength" or frequency of all types of cycles of which all others are but harmonics or derivations from it.

---

[17] I.e., the Foundation's private magazine.
[18] Ray Tomes, "Towards a Unified Theory of Cycles," paper presented at The Foundation for the Study of Cycles conference proceedings February 1990, www.cyclesresearchinstitute.org, p. 4, italicized emphasis added, boldface emphasis in the original.

c. The Well-Tempered Clavier: The First Physical Unification

The suggestion is not as far removed from physical reality as it might at first seem, for the modern Western musical scale with its 12 equidistant "notes" are in fact the example of the first unification in physics. In order to see how, one may perform a simple exercise. If one sits at an acoustic piano and silently presses the note "C," and then hits the same note "C" an octave lower, one will hear the strings of the silently pressed note vibrating sympathetically with the struck note. The reason is simplicity itself. Each string on the piano vibrates not only with the entire length of the string, but simultaneously also vibrates in various *fractions* of that length. Thus, each note has an "overtone" or "harmonic" series of notes. Therefore, one can then press silently the next note in "C's" harmonic series, the note "G," and hit the same "C" as before. Once again, one will hear the silently pressed note "G" vibrating sympathetically with the struck "C." The next note in the harmonic series above "G" is again the note "C," then the note "E," and so on. If one is sitting at the piano keyboard performing this simple experiment, one will notice that the *intervals* of each harmonic overtone of the original "C" are growing *shorter*, first the octave (the first silently pressed "C"), then a *fifth*, the note "G," then again another "C" above that, which is a *fourth*, then the note "E," which is a *third*, and so on. But eventually one will arrive, in the *naturally occurring* harmonic series, at a note that lies somewhere in the "crack" between the notes "A" and "B flat" on the piano keyboard.

But why does the piano (or any other keyboard instrument, for that matter) not have that note? The answer is simple. If that note were present, then it would be *impossible* to play a piece that continually changed keys. One would only be able to play a very limited series of chords. To change keys from, say, "C" to "D," one would literally have to stop and retune the whole keyboard. So what has happened?

What happened was that between the stylistic change in music between the Renaissance and the Baroque, musicians learned to "tweak" the harmonic series, to *tamper* with it, or, as they liked to say, to "*temper*" it, by a slight mathematical adjustment of the natural overtone series, that would in turn create 12 equidistant notes, *each one a harmonic of all others,* and thus music could change through as many keys as it wished in the course of a piece, without having to stop and retune the whole instrument. In this way, each harmonic series of each note on the keyboard, which originally and naturally did *not* overlap completely, were engineered to do so, and thus were unified.[19]

In short, and bearing this analogy in mind, what physicist David Bohm is actually suggesting, as a way forward into the deeper physics of engineering

---

19 For the subject of the *engineering* approach to physics unification, rather than a *theoretical* approach to such unification, see my *The Giza Death Star* (Adventures Unlimited Press, 2001).

the medium itself, is that there is a frequency *of which **all** others — from sound to electromagnetic waves even to gravitational waves — are harmonics*. He is proposing, in effect, a very ancient idea, that of the music of the spheres, a well-tempered clavier of the universe itself. He is proposing a modern physics analogue to the "lost word" and "lost chord" of esoteric lore.

d. Nikolai Kozyrev's Causal Mechanics and Precursor Engineering

If David Bohm clearly implied a direct engineerability of the physical medium along the lines of a "well-tempered" harmonic unification of physics, then Russian astrophysicist Dr. Nikolai Kozyrev went one step further by implying the ability to engineer not *effects*, but *causes* directly.

Such an ability implies that Kozyrev had a rather more unconventional view of time than does standard physics. For Kozyrev, time was not a mere duration, a passive stage on which physical events were enacted;[20] rather, time itself was an actor on the stage that, like space, had a multi-dimensional character and *quality*. An analogy will be useful to illustrate this point. Ordinarily physics tends to think of time in terms of the future, the present, and the past, i.e., as simple *duration*.[21] But human natural languages view time in a much subtler, deeper, and "qualitative" sense, with a variety of verb tenses and voices: future perfect, pluperfect, past perfect, active and passive voices, and so on. In a sense, natural languages therefore view time and the correlations between systems in a much deeper and more sophisticated manner than does physics.

It was this subtlety that Kozyrev intended to explore, and to render into the formally explicit language of mathematics. This complexity and subtlety of interrelationships Kozyrev located in the "rotation moment" of a given system, that is, the subtleties of time and of the interactions of systems could be modeled as a series of interlocking, and *interfering*, systems of rotation, or "dynamic torsion."[22] Time itself could impart its own intensity — Kozyrev's word for "compression," implying its opposite, rarefaction — to a system, as well as impart a spin orientation to a system.[23] Through a series of extremely subtle experiments with gyroscopes, balances, and in some cases, even telescopes, Kozyrev was able to determine that prior to the inception of any physical action, a kind of "pre-action" would be recorded by his measuring equipment, almost as if the equipment was "anticipating" the physical action

---

20 Q.v. my *The Philosophers' Stone* (Feral House, 2009), pp. 164–65.
21 Ibid., pp. 151-169.
22 Farrell, *The Philosophers' Stone*, pp. 166-167.
23 Ibid., pp. 176-178.

itself.[24] "Cause" and "effect" were therefore themselves the result of a lower-dimensional fracture or breaking of a higher-dimensional unity and symmetry, much like Bohm's projection analogy. By noting the temporal conditions under which these "pre-actions" or "precursors" occurred, one could eventually *engineer* the precursors to any physical action.

Thus, bearing in mind Kozyrev's "precursor engineering," and to return to Bohm's analysis of matter itself as a set of waves in the medium that are interfered upon one another, producing what Dewey and Dakin would call an "averaged" wave, then one may imagine a horrific possibility, namely, the exact mirror image of such a wave which, interfered upon the original, would sum to zero, or exactly cancel it out, making the "something" a nothing again. The ultimate in "precursor" engineering was the power to erase an effect of a physical action altogether, by eradicating its cause.

Furthermore, Bohm's and Kozryev's deeper physics behind the apparently stochastic processes of quantum mechanics implies a similar deeper physics behind the apparently stochastic processes of econophysics as well. Indeed, if one can, *pace* Kozyrev, engineer the precursors of effects — if one can actually engineer *causes* — then this implies that one could indeed engineer the precursors of *economic* activity, since the means to do so, direct engineering of the physical medium itself, is implied in both cases.

Consequently, one now has a speculative basis upon which to advance the reasons for the close association of the banking class throughout history with the temple, that is to say, with that element or class within human society that has at least *some* grasp on this deeper physics: banking is in effect an alchemical operation of creating information out of nothing, in this case the information of credit and debt, for the latter is but a dim and pale technological reflection in the realm of finance of the analogous, but deeper, operations in physics. And it is similarly likely that this relationship was formed precisely so that by utilizing the latter, the former technology and deeper physics might ultimately be recovered. In short, were such a physics to be recovered, an alliance with as many temple priesthoods was altogether necessary, for each most likely preserved some fragment of it, which when appropriately assembled would be once more accessible. It provides also a rationale for their obsessive interest, since ancient times, in "world unification and domination," for on the one hand, it is likely that vast financial resources — resources transcending the wealth of any one civilization in ancient or modern times — would be needed to reconstruct such a deeper physics, and on the other hand, a world extent is needed in order to maintain the suppression of any independent development or recovery of such a technology by potential rivals.

---

24   Ibid., p. 179.

B. Economics, Astrology, and Astrophysics

There are two streams of data that now converge to exhibit a probable deeper physics to economic activity: on the one hand, the vast database of wave-like forms from the Foundation for the Study of Cycles, and on the other, the implications of David Bohm's implicate order, where matter itself is the result, and hence a natural resonator, of such waves. As noted in chapter two, one of the implications of the Foundation for the Study of Cycles' database was that economic cycles, precisely because they seemed to have some sort of underlying physics basis — recall Dewey's and Dakin's sound wave analogy once again — were to that extent and in a certain sense "beyond the total rule of man's conscious will."[25] But there is another kind of database that Dewey and Dakin did *not* consult, but it too suggests a profoundly deep physics to economic activity.

And that database is astrology.

Astrologers have, of course, been casting "mundane horoscopes" for various nations for decades, if not centuries and millennia, and more recently, many have noticed the odd correlations of certain recurrent planetary alignments and periods of economic boom or bust. One to do so is Robert Gover in a recent, and fascinating, book called *Time and Money: The Economy and the Planets*. He begins by noting the fact that Saturn orbits the sun every 28–30 earth years, Uranus 84 years, Neptune 165 years, and Pluto 248 years.[26] He then notes the importance of these outer planets for astrological economic observation: since the other planets — Jupiter, Mars, the Earth, Venus, and Mercury — all move too rapidly around the Sun, they cannot be used "to mark years or decades of major economic cycles."[27] He then states his main thesis, one more or less well-known to astrologers, but not well-known outside such circles:

> Every time the USA has gone through a great depression, the outermost, slowest moving planets have formed what astrologers call a grand cross with the USA's natal Sun and Saturn. Every time Uranus has returned to early Gemini where it was when the USA was "born" July 4, 1776, America has experienced its worst wars. Every time Uranus and Pluto have moved into conjunctions or 90-degree squares and simultaneously come conjunct, opposite of square sensitive points in the U.S. birth chart, America has experienced social changes or upheavals....

---

25 Dewey and Dakin, *Cycles: The Science of Prediction*, p. 191.
26 Robert Gover, *Time and Money: The Economy and the Planets* (Hopewell Publications, 2005), p. 2.
27 Ibid.

Other wars occur when the U.S. natal Uranus is "afflicted" by transiting planets, as happened when the World Trade Center and Pentagon were attacked. Saturn and Pluto form 180-degree oppositions three times a century, the latest being in effect on September 11, 2001;. The previous Saturn-Pluto opposition coincided with the tempestuous period we now call the Sixties, the one before that with what we now call The Great Depression.[28]

Gover then explains what four of the most significant conjunctions or alignments are. We shall focus on only two of them:

> A *Grand Cross* aspect is created when four planets form simultaneous squares and oppositions to each other. This is a rare aspect which brings obstructions, tensions, frustrations, i.e. a grand cross to the U.S. Sun-Saturn square has formed each time the USA has fallen into a great depression.
>
> A *Grand Trine* aspect is also rare, and is formed by a triangle of three planets 120 degrees from each other, creating harmonious flows of energy, good fortune and opportunities.[29]

It is odd that the Grand Cross conjunction bears such a strange resemblance to the planetary positions charts of RCA engineer Nelson.

Grand Crosses form such a common feature to U.S. depressions that Gover actually formulates it as a kind of astrological law: "No grand cross, no great depression."[30] However, Gover notes that in this respect, astrological "prediction" is not to be misconstrued:

> If we view the planets around us in our solar system as a huge celestial clock, the first thing history teaches us is that the celestial clock is not mechanically precise like our earthly clocks. Although we can discern from history when like economic events are due, clock-like prediction isn't possible. *Certain planetary patterns create seasons when certain types of events can be expected.* But the planets cannot tell us specifically how events will unfold, nor how we will respond. We know when winter is nigh but not how cold it will get. Some hurricane seasons bring great devastation, others are less severe.[31]

---

28 Gover, *Time and Money*, pp. 2–3.
29 Ibid., p. 10, emphases in the original.
30 Ibid., p. 51.
31 Gover, *Time and Money*, p. 70, emphasis added.

With this said, a glance at Gover's charts is in order.

The first American Great Depression occurred in the 1780s. His chart looks like this:

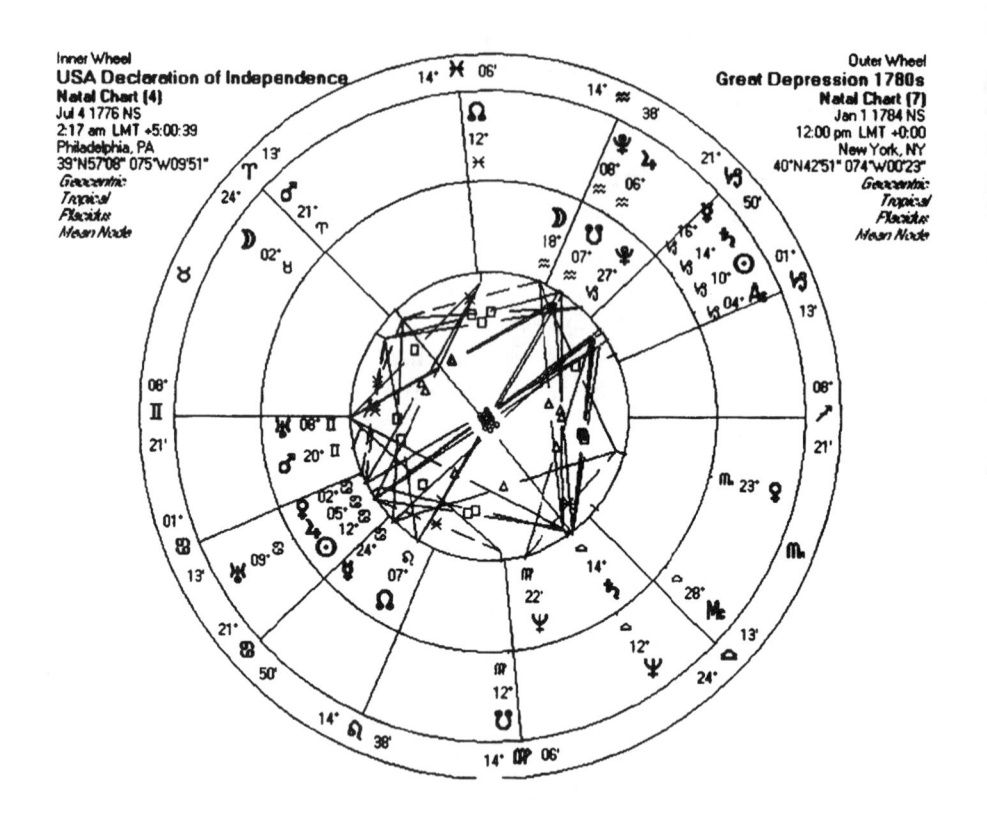

Robert Gover's Astrological Chart for the American Great Depression of the 1780s [32]

If one takes the time to decipher the planetary symbols, then one finds "Saturn was at 15 degrees Capricorn which puts it opposite the U.S. natal Sun at 13 Cancer; meanwhile, Mars at 21 Aries is within orb or an opposition to (the U.S.A.'s) natal Saturn at 15 degrees of Libra."[33] In short, a Grand Cross was formed between the planetary positions during the 1780s great depression, and that of their positions at the time of the U.S.A.'s birth.

Similarly, a Grand Cross is formed during the Great Depression of the 1870s.

---

32   Ibid., p. 41.
33   Ibid., p. 40.

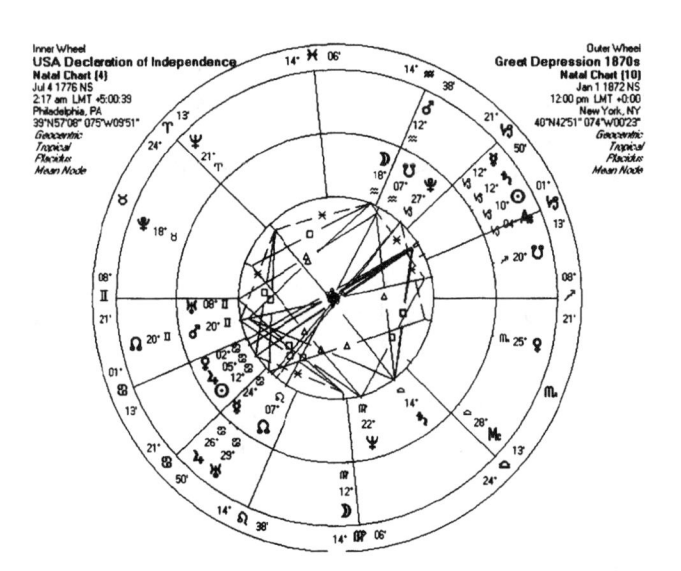

Robert Gover's Astrological Chart for the American Great Depression of the 1870s [34]

During this time "Saturn in Capricorn formed a square with Neptune in Aries to create the grand cross with (the U.S.A.'s) Sun and Saturn."[35]

Finally, during the Great Depression of the 1930s, yet another Grand Cross was formed:

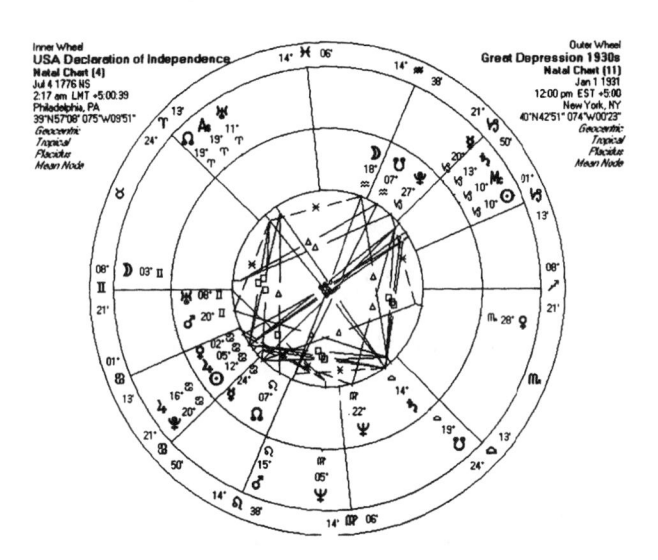

Robert Gover's Astrological Chart for the American Great Depression of the 1930s [36]

---

34 Gover, *Time and Money*, p. 44.
35 Ibid.
36 Gover, *Time and Money*, p. 45.

This chart chows a Grand Cross between Saturn, at 13 degrees Capricorn square with Uranus at 11 degrees Aries, and opposite (i.e., at 180 degrees) with the U.S.A.'s natal Sun and Saturn.[37]

That these Grand Crosses have a malign influence would seem, at least as a *prima facie* case, to be a given. But is there any real-world physics correlation?

The answer to this question is as easy as reproducing one of Nelson's charts from his RCA study of planetary alignments and radio signal propagation.

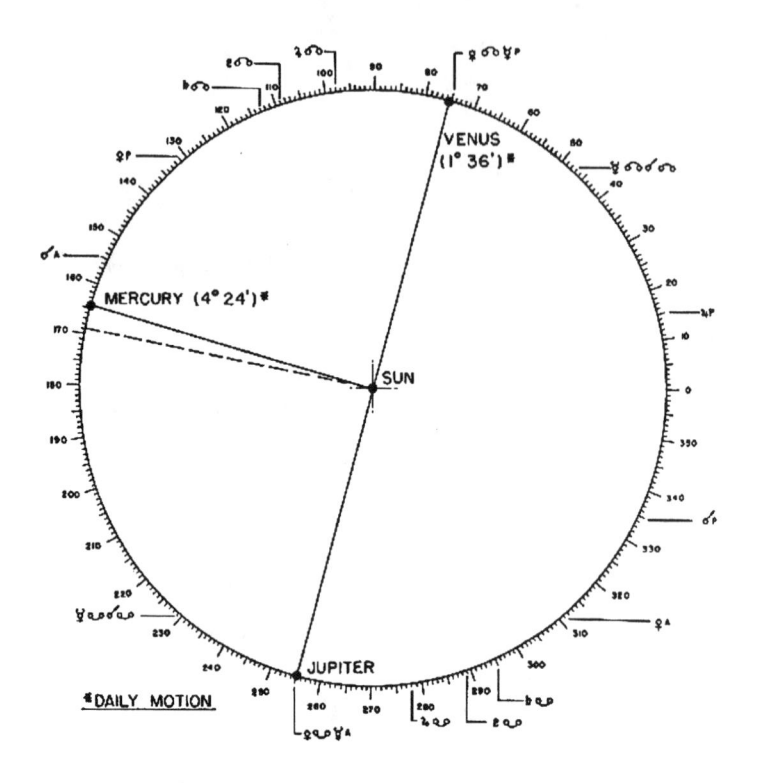

Nelson's Chart of Venus-Jupiter Opposition [38]

Note the similarity is once again that planets, in this case the *inner* planets, are in certain relationships to each other, relations of 90 degrees or some harmonic thereof, i.e., 180 or 270 degrees.[39]

---

37 Ibid., p. 47.
38 James Nelson, "Shortwave Radio Propagation Correlation With Planetary Positions." Conference paper presented to the AIEE Subcommittee on Energy Sources, AIEE General Winter Meeting, January 1952.
39 Ibid. The *dissimilarities* should also be noted, for Gover's focus is on the outer planets, namely, the gas giants and Pluto (recently demoted from planetary status). Additionally, Gover, in classical astrological fashion, notes which zodiacal house the planets are in, whereas Nelson does not.

Thus, if the Foundation for the Study of Cycles' database and Gover's (and other astrologers') charts are any indicator, then we can now draw even closer to an understanding of the relationship that was seen to exist between banking and the temple in ancient times, for *the astrological data of economic boom and bust would have been known by them from ancient times, and it would have been crucial for the "financial powers in the know" to have such data available, in order to exacerbate or damp the overall trend of boom or bust within a cycle.*

With this in mind, a glance at the evidence gathered by Ellen Hodgson Brown, whose book was mentioned in the Prologue, is in order, for a comparison of the activities of banksters relative to Gover's astrological charts is quite revealing.

C. Ellen Hodgson Brown
1. The Depression of the 1780s and the Banksters

Brown points out something that most modern Americans do not know, and that is, that prior to the American Revolution, most of the colonies printed their own paper money — *debt-free* — and actually made loans to farmers and businessmen. The result was a booming economy and almost full employment. When Benjamin Franklin went to England prior to the revolution, he was asked about the source of this prosperity "by the directors of the Bank of England," and Franklin responded that the colonies "issued paper money 'in proper proportion to the demands of trade and industry.'"[40] But what was the "backing" of this money? The colonies, however, had little silver and gold with which to back their issues of paper currency. With what, then, was it backed? The then famous Protestant minister in New England, Cotton Mather, made clear what the backing of this colonial scrip was by asking a pointed series of questions:

> Is a Bond or Bill of Exchange for (one thousand pounds), other than paper? And yet it is not as valuable as so much Silver or Gold, supposing the security of Payment is sufficient? *Now what is the security of your Paper-money less than the Credit of the whole Country?*[41]

As Brown notes, "Mather had redefined money. What it represented was not a

---

40 Ellen Hodgson Brown, *Web of Debt: The Shocking Truth about Our Money System and How We Can Break Free* (Baton Rouge, Louisiana: Third Millennium, 2008), p. 39. If there is one book the reader should read about the history of privately-issued debt money versus state-issued credit money, this book is it. It is thoroughly documented and well-written. Brown especially excels at exposing the fallacious assumptions of many that a mere return to the "gold standard" would resolve the problem. As we have already seen, bullion itself is subject to manipulation by the "bullion brokers" or banksters.

41 Brown, op. cit., p. 37.

sum of gold or silver. It was credit: 'the credit of the whole country.'"[42] Within the context of the evidence presented thus far, and in particular, the analysis of David Astle reviewed in chapter five, what Mather had *really* done is return to the very ancient conception of money prior to the rise of the international bankster class of bullion brokers in ancient times; what he had done was to return to the idea of money as a *credit bill* against the surpluses of the state warehouse, and not *an interest-bearing note of private issuance*.

Franklin stated this conception somewhat differently: "The riches of a country are to be valued by the quantity of labor its inhabitants are able to purchase and not by the quantity of gold and silver they possess."[43] The difference is striking, for

> when gold was the medium of exchange, money determined production rather than production determining the money supply. When gold was plentiful, things got produced. When it was scarce, men were out of work and people knew want. *The virtue of government-issued paper scrip was that it could grow along with productivity, allowing potential to become real wealth.* [44]

Franklin elaborated on the source of colonial prosperity to his English hosts, and his words are worth taking to heart, for in them one discerns the clear difference between a closed system of "debt-as-money" or monetized debt, and an open system of money as a medium of exchange of *production and credit:*

> In the colonies we issue our own money. It is called Colonial Scrip. We issue it to pay the government's approved expenses and charities. We make sure it is issued in proper proportions to make the goods pass easily from the producers to the consumers…. In this manner, creating for ourselves our own paper money, we control its purchasing power, and *we have no interest to pay to anyone.* You see, a legitimate government can **both spend and lend money into circulation, while banks can only lend significant amounts of their promissory bank notes,** for they can neither give away nor spend but a tiny fraction of the money people need. Thus, when your banks here in England place money in circulation, there is always a debt principal to be returned and usury to be paid. The result is that you have always too little credit in circulation to give the workers full employment. *You do not have too many workers, you have too little money in circulation,*

---

42   Ibid.
43   Ibid.
44   Ibid., emphasis added.

and that which circulates, all bears the endless burden of unpayable debt and usury. [45]

Franklin has seen the essential criminality and fraud that is central banking, for the governments pursuing a policy of "monetizing the debt" only means that they are beholden to a private monopoly, which issues "debt as money," whereas the colonial experience — and the very *ancient* experience — was that *true* money was *credit on the productive surplus of the state*, and hence, *only the state could issue it.*

Needless to say, England's banksters were not about to allow this situation to continue, allowing the colonists to gain prosperity without enriching their own parasitic coffers. Thus, the Bank of England parlayed its influence in Parliament to get the 1764 Currency Act passed, which made it illegal for the colonies to issue their own money. And predictably, as Franklin observed, a year later the streets of the colonies were filled with the unemployed and beggars.[46] And it was this substitution of debt as money, the replacement of *real* money by the *facsimile* of money that, according to Franklin, was the real cause of the Revolution.[47]

When the Revolutionary War finally came, the Continental Congress financed the entire endeavor by once again resorting to the expedient of issuing its own paper scrip as a circulating debt note of the state to be redeemed by coinage at a future date.[48] Of course, the Continental Congress issued too much of this scrip, some two hundred million dollars' worth, so that by the conclusion of the Revolutionary War, the scrip was basically worthless.

But the real lesson was not in the dangers of a state hyper-inflating its state-issued credit notes. The real lesson was that the Continental Congress' scrip

> Still evoked the wonder and admiration of foreign observers, because it allowed the colonists to do something that had never been done before. They succeeded in financing a war against a major power, with virtually no "hard" currency of their own, *without taxing the people.* Franklin wrote…during the war, "the whole is a mystery even the politicians, how we could pay with paper that had no previously fixed fund appropriated specifically to redeem it. *This currency as we manage it is a wonderful machine.* Thomas Paine called it a "corner stone" of the Revolution: "Every stone in the Bridge, that has carried

---

45 Brown, op. cit, pp. 40–41, italicized emphasis in the original, boldface emphasis added.
46 Ibid., p. 41.
47 Ibid.
48 Brown, op. cit., p. 43.

us over, seems to have claim upon our esteem. But this was a corner stone, and its usefulness cannot be forgotten."[49]

Alas, it is a lesson the American people and the two political parties that supposedly "represent" them seems all but forgotten in modern times.

Of course, the British were fully aware of how their rebellious colonies were funding their Revolution, and purposed to crash the currency by the time-tested tactic of counterfeiting. One British general cited by Brown in her book noted that every art of counterfeiting had been tried, but, to his chagrin, "still the currency has not failed."[50] It was only after the successful Revolution that the Continental Scrip failed, as the very same Founding Fathers grew understandably disillusioned with the resulting devaluation that inflation of the supply — not to mention the counterfeit scrip in circulation — caused. The result of this deliberate speculation against the American Continental Scrip was predictable, for the Founding Fathers rebelled against the very paper money-as-credit against the state's *future* productive surplus after the war by stating that Congress had the power to make "and *coin*" money. In other words, the Continental Congress had fallen into the old trap of the issuance of the *facsimile of money*, of money as *debt*, even if that debt were the credit against a future surplus of the state.

> The notes represented debt, and the debt had now come due. The bearers expected to get their gold, and the gold was not to be had. There was insufficient supply of money for conducting trade. Tightening the money supply by limiting it to coins had quickly precipitated another depression. In 1786, a farmers' rebellion broke out in Massachusetts, led by Daniel Shays. Farmers brandishing pitchforks complained of going heavily into debt when paper money was plentiful. *When it was no longer available and debts had to be repaid in the much scarcer "hard" coin of the British bankers, some farmer lost their farms.* [51]

The immediate result of this First American "Great Depression" then, was of course the call for a stronger central government and a means for it to create "an expandable money supply," and the convocation of the assembly that eventually led to the draft of the current American Constitution.[52]

At this juncture, it is worth citing Ellen Brown's comments extensively:

---

49 Ibid., emphasis in the original.
50 Ibid., p. 44.
51 Brown, op. cit., p. 47, emphasis added.
52 Ibid.

The solution of Treasury Secretary Hamilton was to "monetize" the national debt, by turning it into a source of money for the country. He proposed that a national bank be authorized to print up banknotes and swap them for the government's bonds. The government would pay regular interest on the debt, using import duties and money from the sale of public land. Opponents said that acknowledging the government's debt at face value would unfairly reward the speculators who had bought up the country's I.O.U.s for a pittance from the soldiers, farmers and small businessmen who had actually earned them; but Hamilton argued that the speculators had earned this windfall for their "faith in the country." He thought the government needed to enlist the support of the speculators, *or they would do to the new country's money what they had done to the Continental....*

Jefferson, Hamilton's chief political opponent, feared that giving private wealthy citizens an ownership interest in the bank would link their interests **too** closely within it. The government would be turned into an oligarchy, a government of the rich at war with the working classes. A bank owned by private stockholders, whose driving motive was profit, would be less likely to be responsive to the needs of the public than one that was owned by the public and subject to public oversight. Stockholders of a private bank would make their financial decisions behind closed doors, without public knowledge or control.

But Hamilton's plan had other strategic advantages, and it won the day. Besides neatly disposing of a crippling federal debt and winning over the "men of wealth," it secured the loyalty of the individual States by making their debts too exchangeable for stock in the new Bank. The move was controversial; but by stabilizing the States' shaky finances, Hamilton got the States on board, thwarting the plans of the pro-British faction that hoped to split them up and establish a Northern Confederacy.[53]

It is worth pausing at this juncture to observe, in the context of Gover's astrological chart of the Great Depression of the 1780s, what we have:

1) The Continental Congress' scrip was essentially *not* the same thing as Colonial scrip, in that it *was* debt money issued against a *future* promise to pay, some of which was a promise to pay in bullion which the colonies did *not* have in abundance. Hence, the Continental Scrip inevitably opened itself to speculation and

---

53 Brown, op. cit., pp. 47–48, emphasis added.

counterfeiting by the very European and British banksters who held a virtual monopoly on bullion supplies;
2) The inevitable result, as the value of the Continental plunged after the Revolution and the money supply contracted to reflect the scarce bullion supply, was that the post-Revolutionary states could not pay their debt, and an inevitable depression occurred as the supply of money contracted and private debt-holders were not able to service that debt;
3) the result of *this* sequence of events led to the calling of the Constitutional convention, the formulation of the current American constitutional system, and the first chartered private central bank of the United States which issued the facsimile of money based on monetized debt;
4) The alternative fear, which Hamilton voiced, was that "private speculators" would manipulate any new American currency via speculation and counterfeiting, driving it to similar worthlessness, and splitting the new American nation if they themselves were not given some stake in the new currency as a vested class interest.

In short, almost from the beginning of the current constitutional system of America, an uneasy compromise — a *détente* — was struck with the banksters to allow the new nation to survive, and enrich that class in the process. And during all this period, as Gover noted, the planets were in certain alignments....

2. *The Depression of the 1870s and the Banksters*

The American depression of the 1870s again followed yet another war in American history, the War Between the States. And like the American Revolution, at least one of the leaders of the belligerent parties, as is now well-known, chose to finance his side of the war by issuing state-created debt-free money, i.e., money as *credit on the productive output of the nation*. His name, of course, was Abraham Lincoln, and the fiscal lessons of his presidency and its immediate aftermath are once again worth rehearsing in some detail.

German Chancellor Otto Von Bismarck wrote a curious thing in 1876 about the fiscal policies of the Lincoln Administration:

> I know of absolute certainty, that the division of the United States into federations of equal force was decided long before the Civil War by the high financial powers of Europe. These bankers were afraid that the United States, if they remained in one block and

as one nation, would attain economic and financial independence, which would upset their financial domination over Europe and the world. Of course, in the 'inner circle' of Finance, the voice of the Rothschilds prevailed. They saw an opportunity for prodigious booty if they could substitute two feeble democracies burdened with debt to the financiers... in place of a vigorous Republic sufficient unto herself. Therefore, they sent their emissaries into the field to exploit the question of slavery and to drive a wedge between the two parts of the Union.... *The rupture between the North and the South became inevitable; the masters of European finance employed all their forces to bring it about and to turn it to their advantage.* [54]

There was just one problem. President Lincoln refused to go into debt to the private class of banksters to fund the Northern effort in the Civil War. Chancellor Bismarck's comment is worth citing:

> The Government and the nation escaped the plots of the foreign financiers. They understood at once, that the United States would escape their grip. The death of Lincoln was resolved upon.[55]

Bismarck, in other words, in his customary direct way, was simply stating that the "inner circle of European financiers" led by the Rothschilds had had Lincoln murdered as a punishment and message to those who dare presume to challenge their power.

While Lincoln was busily issuing his Greenback debt-free scrip, however, the banksters were busily hatching a scheme of their own in Congress through their own faction. Once again, it is essential to cite Ellen Brown extensively in order to appreciate what the scam was, and how it was effected:

> While one faction in Congress was busy getting the Greenbacks issued to fund the war, another faction was preparing a National Banking Act that would deliver a monopoly over the power to create the nation's money supply to the Wall Street bankers and their European affiliates. The National Banking Act was promoted as establishing safeguards for the new national banking system; but while it was an important first step toward a truly national bank, it was only a compromise with the bankers, and buried in the fine print, it gave them exactly what they wanted. A private communication from a Rothschild investment house in London to an associate banking firm

---

54 Brown, op. cit., pp. 89–90, emphasis added by Brown.
55 Ibid., p. 91.

in New York dated June 25, 1863, confided:

"The few who understand the system will either be so interested in its profits or so dependent upon its favors that there will be no opposition from that class, while, on the other hand, the great body of people, mentally incapable of comprehending... will bear its burdens without complaint."

The Act looked good on its face. It established a Comptroller of the Currency, whose authority was required before a National Banking Association could start business. It laid down regulations covering minimum capitalization, reserve requirements, bad debts, and reporting. The Comptroller could at any time appoint investigators to look into the affairs of any national bank. Every bank director had to be an American citizen, and three-quarters of the directors had to be residents of the state in which the bank did business. Interest rates were limited by State usury laws; and if no laws were in effect, then to 7 percent. Banks could not hold real estate for more than five years, except for bank buildings. National banks were not allowed to circulate notes they printed themselves. Instead, they had to deposit U.S. bonds with the Treasury in a sum equal to at least one-third of their capital. They got government-printed notes in return.

So what was the problem? Although the new national banknotes were technically issued by the Comptroller of the Currency, this was just a formality, like the printing of Federal Reserve Notes by the Bureau of Engraving and Printing today. The currency bore the name of the bank posting the bonds, and it was issued at the bank's request. *In effect, the National Banking Act authorized the bankers to issue and lend their own paper money.* The banks "deposited" the bonds with the Treasury, but they still owned the bonds; *and they immediately got their money back in the form of their own banknotes.* Topping it off, the National Banking Act effectively removed the competition to these banknotes. It imposed a heavy tax on the notes of the state-chartered banks, essentially abolishing them. It also curtailed competition from the Greenbacks, which were limited to specific issues while the bankers' notes could be issued at will. *Treasury Secretary Salmon P. Chase and others complained that the bankers were buying up the Greenbacks with their own banknotes.* [56]

In other words, what the National Banking Act really did was allow the banksters effectively to outproduce the government Greenbacks in issues

---

56   Brown, op. cit., pp. 91–92, emphasis added.

of their own debt-bearing banknotes, which the banksters then used to buy Greenbacks and take them out of circulation!

The banksters followed this up in 1873 with the so-called Act that became known popularly as the "Crime of '73," an act which effectively forbade the coinage of silver as legal tender, effectively placing America on the gold standard once again. Predictably, the Act led to a vastly shrunken money supply, unemployment, and the Depression of the 1870s. The result was similar to the bankster-engineered Depression of the 1780s, for it led to a political "revolt" of powerful farmers who formed the appropriately named Greenback Party, calling for the issuance of state-created debt-free money directly, which money was to be used putting people back to work improving the infrastructure of the country.[57] While the Greenbacks never succeeded in placing a national candidate of their own into the White House, their message was heard, and in 1881 James Garfield became President of the United States of America. Garfield proclaimed:

> Whosoever controls the volume of money in any country is the absolute master of all industry and commerce...And when you realize that the entire system is very easily controlled, one way or another, by a few powerful men at the top, *you will not have to be told how periods of inflation and depression originate.* [58]

As Brown notes, "Garfield was murdered not long after releasing this statement, when he was less than four months into his presidency."[59] We shall have occasion to return to a consideration of President Garfield's remarks in a moment, but for now, be it noted, that he has clearly insinuated that the cycles of boom and bust are *artificially created by the banksters, whereas the cycles data assembled by Edward Dewey, and the astrological data assembled by Gover, clearly indicate that such things appear to have much deeper causes than just human actions, and are, to a certain extent, inevitable. So....what is really going on?*

A hint is perhaps gained by the fact that

1) in the cases of both depressions examined thus far, both occurred *after* a major war; and,
2) both occurred after the government in each case decided to fund the war by issuance of debt-free money, bypassing the banksters completely; and,

---

57 See Brown's discussion on pp. 93–95.
58 Ibid., p. 94, emphasis added.
59 Ibid.

3) the banksters retaliated against that currency by unleashing various forms of speculation and manipulation against it, through counterfeiting, or by otherwise removing the government's currency from circulation; and
4) in each instance thus far, this resulted in the deliberate tightening of the money supply and a corresponding loss of jobs, production, and an economic depression.

These patterns reach the nadir of their expression in the greatest Depression of them all, the Great Depression of the 1930s.

3. The Great Depression of the 1930s and the Banksters

By the time of the Great Depression, the great struggle between the federal government of the United States of America and the private banksters had finally been won by the latter with the creation of the privately owned Federal Reserve Bank in 1913, and its thinly-disguised police force agency, the Internal Revenue Service, the agency responsible for gathering the newly created federal income tax that was designed specifically to pay the regular interest payments to the banksters loaning money at interest to the government. Then began the deliberately and quietly orchestrated run-up to the Great Depression:

> The problem began in the Roaring Twenties, when the Fed made money plentiful by keeping interest rates low. Money seemed to be plentiful, but what was actually flowing freely was "credit" or "debt." Production was up more than wages were up, so more goods were available than the money to pay for them; but people could borrow. By the end of the 1920s, major consumer purchases such as cars and radios (which were then large pieces of furniture that sat on the floor) were bought mainly on credit. Money was so easy to get that people were borrowing just to invest, taking out short-term, low-interest loans that were readily available from the banks.
>
> The stock market held little interest for most people until the Robber Barons started promoting it, after amassing large stock holdings very cheaply themselves. They sold the public on the idea that it was possible to get rich quick by buying stock on "margin" (or on credit). The investor could put a down payment on the stock and pay off the balance after its price went up, reaping a hefty profit. This investment strategy turned the stock market into a speculative pyramid scheme, in which most of the money invested did not actually exist....

The public went wild over this scheme. In a speculative fever, many people literally "bet the farm."... Homesteads that had been owned free and clear were mortgaged to the bankers, who fanned the fever by offering favorable credit terms and interest rates. The Federal Reserve made these favorable terms possible by substantially lowering the discount rate — the interest rate member banks paid to borrow from the Fed. The Fed thus made it easy for the banks to acquire additional reserves, against which they could expand the money supply by many multiples with loans.[60]

But why, asks Ellen Brown, would the Federal Reserve want to swamp the U.S. economy with an inflated supply of borrowed "dollars" of Federal Reserve notes?

The answer is chilling.

"The evidence," says Brown, "points to a scheme between Benjamin Strong, then Governor of the Federal Reserve Bank of New York, and Montagu Norman, head of the Bank of England, to deliver control of the financial systems of the world to a small group of private central bankers."[61] The reason, according to Dr. Carroll Quigley of the Georgetown School of International Relations, and himself an "insider" with access to the banksters' plans, was that during the 1920s, the privately-owned central banks were determined "to use the financial power of Britain and the United States to force all the major countries of the world to go on the gold standard and to operate it through central banks free from all political control..."[62] In other words, the scheme, by forcing money to be a reflection of the world's gold supplies, was one of *drastically curtailing the amount of money in circulation as debt,* thus setting off a Depression.

The plan was indeed ingenious for its cold calculation and cunning:

Norman, as head of the Bank of England, was determined to keep the British pound convertible to gold at pre- [World War I] levels, although the pound had lost substantial value as against gold during World War I. The result was a major drain of British gold reserves. To keep gold from flowing out of England into the United States, the Federal Reserve, led by Strong, supported the Bank of England by keeping U.S. interest rates low, inflating the U.S. dollar. The higher interest rates in London made it a more attractive place for investors to put their gold, drawing it from the United States to England; but

---

60 Brown, op. cit., p. 140.
61 Ibid.
62 Quigley, *Tragedy and Hope,* p. 326.

the lower rates in the United States caused an inflation bubble, which soon got out of hand. The meetings between Norman and Strong were very secretive, *but the evidence suggests that in February 1929,* **they concluded that a collapse in the market was inevitable and that the best course was to let it correct "naturally"** *(naturally, that is, with a little help from the Fed).* They sent advisory warnings to lists of preferred customers…telling them to get out of the market. *Then the Fed began selling government securities in the open market, reducing the money supply by reducing the reserves available for backing them. The bank-loan rate was also increased, causing rates on brokers' loans to jump to 20 percent.*

The result was a huge liquidity squeeze — a lack of available money. Short-term loans suddenly became available only at much higher interest rates, making buying stock on margin much less attractive. As fewer people bought, stock prices fell, removing the incentive for new buyers to purchase stocks bought by earlier buyers on margin. Many investors were forced to sell at a loss by "margin calls" (calls by brokers for investors to bring the cash in their margin accounts up to a certain level after the value of their stocks had fallen). The panic was on, as investors rushed to dump their stocks for whatever they could get for them. The stock market crashed overnight. People withdrew their savings from the banks and foreigners withdrew their gold, further depleting the reserves on which the money stock was built….It was dramatic evidence of the dangers of delegating the power to control the money supply to a single autocratic head of an autonomous agency.[63]

Once again, one has, if one compares this evidence with Gover, that as the planets entered certain alignments, certain policies were pursued by the banksters, first, to artificially inflate the money supply, extending easy credit and debt, and then suddenly contracting it, allowing stock prices to fall, and allowing the banksters to snap up real assets for a substantially lower price.

Of course, as all of this was going on, Herbert Hoover was President, and as we have already seen, he quietly commissioned Commerce Department economist Edward Dewey to study the reasons for all these Depressions.

4. Implications

So now let us return to the statements of President James Garfield cited previously, and to the questions they raised.

---

63 Brown, op. cit., pp. 141–142, all italicized and boldface emphases added.

Recall that he clearly insinuated that the cycles of boom and bust are *artificially created by the banksters,* whereas the cycles data assembled by Edward Dewey, and the astrological data assembled by Gover, clearly indicate that such things appear to have much deeper causes than just human actions, and are, to a certain extent, inevitable. So....*what is really going on?* In each case thus far examined, as these planetary alignments were occurring, or, more importantly, as a glance at Dewey's cyclic data would indicate, each downturn, and particularly that of the Great Depression of the 1930s, came at a moment when the data suggested that this was more or less *inevitable,* regardless of what actions might have been taken by various governments or banksters.

So let us now make two assumptions, based on the assembled data:

1) The cyclic data of Dewey suggested that a downturn was inevitable precisely during the time of the 1930s Depression, and the cosmological data assembled by Gover suggested a similar inevitability for other depressions in American history; and,
2) The presence, since ancient times, of the close association between the "temple" with its astrological associations and the "banking class" *continued unabated into modern times,* a point readily suggested by the presence of economists of major banks in Dewey's own Foundation. Gover's astrological data, moreover, would be available to any astrologer capable of casting a mundane horoscope, and hence, readily available to anyone so inclined to assemble such charts for comparison with the historical evidences of booms and busts in various economies.

Thus, one is led to an astonished, and rather breathtaking conclusion: it would appear that anyone in possession of such knowledge as Dewey's cycles or astrological data similar to Gover's would be in a position, through careful policy manipulation and the contraction of a money supply based on the facsimile of money-as-circulating-debt, to dramatically exacerbate and capitalize on the overall upward or downward trend of such a cycle. Moreover, *this appears to be exactly what happened in the cases of the three great depressions in American history.*

The exact *mechanism* for these deeper physics influences on human behavior is not here in view, however. It may be the case that certain alignments cause certain *types* of behavior to emerge in the aggregate, which in turn create conditions apt to favor one policy or course of action over another. Or conversely, it may be that certain types of influences block the aggregate ability of humans to perceive the subtle manipulations of these would-be master manipulators, which other types of influences may magnify human perceptual abilities and discretion. We simply do not know how this works. But what we *do* know is that this astrological component and its connection to the banking classes is as old as the civilization with which astrology is associated: Babylon itself. It is an association as old as that of the ancient priesthoods and temples not only with the stars, but with the bullion brokers. It is interesting to note, then, that some allege that the Rothschild family secretly traces its family dynasty back to the Sumerian tyrant Nimrod.[64]

D. Implications of Engineerability: The Ancient Alchemical Connection

All this implies an "engineerability" to economic *trends*, even if one does not possess the technology or means to engineer the physical medium or thus its *cycles themselves*, for if one possessed a sufficient database to know in advance the inevitable cycles of boom and bust, one may not be able to reverse those cycles, but one could considerably exacerbate or alleviate their overall effect.

The keystone in the arch linking together all these disparate concepts — econophysics, astrology, astronomy, Bohm's and Kozyrev's precursor engineering of causation itself, the bullion brokers, the ancient temples — is alchemy, which in its exoteric aspect is the ability to confect the Philosophers' Stone, a substance that can turn base metals into gold, and in its esoteric aspect is the ability to draw upon the transmutative, information-creating properties of the physical medium itself for the power to create or destroy. In its exoteric aspect, therefore, it would be of immediate concern to the bullion brokers to command such a technology, to prevent any outside force or faction from utilizing such a technology to collapse the artificially-created value of their bullion, and hence ruining their private money-creating power. Similarly, it would be of immediate interest to those wishing to break that bullion brokers' monopoly to command that technology. By the same token, the bullion brokers, by controlling the exoteric aspects of alchemy, would see in it a means to the vastly expanded powers of its esoteric aspect, the ability to manipulate and engineer the physical medium itself, thus including even the cycles of aggregate human behavior and activity. Once that power was restored to them, their iron grip on power would be complete.

So notice the stages which have been observed throughout the previous chapters:

---

64 Fritz Springmeir, *Bloodlines of the Illuminati*, p. 237. Springmeir states that "According to their own secret family genealogy, which is recorded in a sacred secret book, the Rothschilds are descended from Nimrod, the great Babylonian warrior leader." Of course, a "secret sacred book" means that there is *no* substantiation of these claims. However, as will be seen in a subsequent chapter, there is a case that can be made of a generalized contextual nature in which such claims may be interpreted, and in which they find a more or less very loose corroboration.

1) In the first stage, the bullion brokers resort to a kind of *false* alchemy, i.e., the private creation of credit (and interest debt) by the fiction of ledger credit entry. This is the financiers' alchemical charlatanism, for something has quite literally been created out of nothing, but that "something" only continues to *be* something of value so long as wider society accredits it as such. It is not genuine information, for it is in fact, in its interest debt creation, the creation of "*negative* information," a black hole of financial entropy that inevitably must suck all creative production into it and let nothing escape; it is a financial cancer that inevitably will consume its host, creating its own death. Lose that confidence, and the value is gone. Thus, the way to sanctify and ensure that "negative value" was to associate their money minting and issuing activities with the sanctity and probity of the various temple priesthoods in each civilization;

2) In the second stage, alchemy is pursued for its own sake, not only to increase or expand the supply of bullion, but also to monopolize that technology lest the too-rapid increase of that supply ruin the value of the privately-created money which they themselves have implemented. In short, they must monopolize this technology lest their money monopoly be challenged by kings intent upon restoring their crown prerogative of money issuance based upon the creative production of their state and subjects. And in order to monopolize it, they must quite literally infiltrate every society and civilization which appears likely to develop it. It is, perforce, and by the nature of the case, an international conspiracy, for it must have an international reach in order to ensure that the economic system remains closed.

3) In the third stage, which is always the ultimate goal, the bullion brokers seek to develop the highest alchemical technology of them all: the ability to engineer the physical medium and its cycles directly. Since alchemy in its *exoteric* aspect is based upon its *esoteric* aspect, i.e., upon the idea of the physical medium as an information-creating and transmutative medium, this technology too must likewise be monopolized, since any rival gaining access to it could successfully challenge their monopoly of money issuance and, more importantly, would have the power to overturn them by force if necessary.

From Morgan's suppression of Tesla, to Tesla's own possible independent use of his wireless impulse transmitter technology in a weaponized fashion at

Tunguska, to Nazi Germany's restoration of state-created, debt-free money and to its investigation of this deeper "torsion"-based physics with its Bell project, to ancient Sparta's and Rome's attempts to restore economic autarchy and Rome's burning of Egyptian alchemical books, the pattern is the same:

Wherever there is a private monopoly on the creation of money, there too one finds the inevitable and covert alliance with the temple of science, and a mutual interest in the secret development of the deep alchemical physics of the medium, for that, far beyond the shadowy imitations of international banksters creating ledger credit entries, is the ultimate power to create or destroy.

That alliance with the temple gave the ancient bullion brokers yet another clue on the road to recovering the lost and unified physics: the placement of such temples on the surface of the Earth itself, and the repetitive occurrence within those structures of a sacred geometry known to its priests and initiates…

III.

THE MONSTERS IN THE MACHINE

"In their drive to advance the global empire, corporations, banks, and governments (collectively the corporatocracy) use their financial and political muscle to ensure that our schools, business, and media support both the fallacious concept and its corollary. They have brought us to a point where our global culture is a monstrous machine that requires exponentially increasing amounts of fuel and maintenance, so much so that in the end it will have consumed everything in sight and will be left with no choice but to devour itself."
—John Perkins, *Confessions of an Economic Hit Man*, p. xv.

Seven

SACRED SITES AND SCALAR TEMPLES
THE EARTH GRID AND THE TRANSMUTATIVE MEDIUM

∴

> *"The modern banking system manufactures money out of nothing. The process is perhaps the most astounding piece of sleight of hand that was ever invented... If you want to be slaves of the bankers, and pay the cost of your own slavery, then let the banks create money."*
> —Lord Josiah Stemp, former Director of the Bank of England[1]

> *"The deep attitude of the global, mostly male, corporate gang was expressed by David Rockefeller at the June 1991 Bilderberg meeting in Baden-Baden, Germany, where he argued for 'supranational sovereignty of an intellectual elite and world bankers, which is surely preferable to the national auto determination practiced in the past centuries.'"*
> —Daniel Estulin, citing David Rockefeller[2]

There is little doubt that many ancient temples — from Giza to Stonehenge to Teotihuacan in Mexico — are aligned to astronomical phenomena or particular stars or constellations. However, it is equally true, as noted in chapter four, that many of these temples are aligned, or rather, situated, to something else, and that is their peculiar siting on the surface of planet Earth herself. Many researchers, noticing this peculiar placement, have been able to demonstrate that these sites, if viewed as a total system, constitute a kind of "gridwork" or imaginary lattice stretched across the face of the globe. Moreover, these structures, from the Great Pyramid of Giza to the Parthenon in Athens, to the

---
1 Cited in Robert Gover, *Time and Money: The Economy and the Planets*, p. 4.
2 Daniel Estulin, *The True Story of the Bilderberg Group*, p. 61.

vast and ancient pyramids of Mexico, appear to incorporate measures in their dimensions comprising constants of a "sacred geometry."

Consequently, in order to grasp the intricate relationships of deep physics and deep finance, one must go yet another step further, for if there was, as has been already seen, a deep and ancient connection between the ancient temples, their priesthoods, and astrology on the one hand and the "bullion brokers" on the other, and if, as has been argued, this connection evinced a hidden motivation on the part of the latter to know and manipulate the physics this portended, then the alignment of these sites and temples, and indeed, their very dimensions, may form yet another clue and data point to be considered. Indeed, researcher Brian Desborough makes some astonishing assertions in this respect:

> Sacred geometry, which comprised the core teaching of any genuine Mystery School of antiquity, *was incorporated into the design of buildings that enabled them to function as resonant cavities, which were attuned to the dynamic energies that pervade the cosmos.* This is the same energy continuum that was harnessed some two thousand years later by Nikola Tesla and other scientific pioneers, who elected to conduct their research beyond the stultifying confines of academia.
>
> The basic tenets of sacred geometry were taught in ancient Mystery Schools, in countries as geographically diverse as India, Britain, and Egypt. *This strongly suggests that such institutions functioned not as individual entities, but rather as a transnational network.* Such a suggestion implies that the more esoterically-oriented ... sects operated not as individual religious communities, but were in communication with the Mystery Schools of other nations.[3]

As has been seen, the rise of an international class of bullion brokers closely associated with such temples in ancient times provides a ready mechanism to explain this transnational character of the Mystery Schools, and may have been instrumental in the dispersion of such ideas.

More importantly, however, Desborough clearly alludes to an occulted *physics purpose* that motivated the incorporation of these geometries into ancient temples, for these were, in his words, designed to function as "resonant cavities" or, in a physicist's language, as "coupled harmonic oscillators" to those "dynamic energies that pervade the cosmos." In this last statement, Desborough is correct, for as has also been seen, the deeper physics suggested by Kozyrev and Bohm in the perpetual dynamism of the cosmos, with its rotating systems within rotating

---

3 Brian Desborough, *They Cast No Shadows: A Collection of Essays on the Illuminati, Revisionist History, and Suppressed Technologies* (San Jose, California: Writers Club Press, 2002), p. 174, emphasis added.

systems, implies that matter itself arises as a gridwork or interference pattern of the waves produced in the physical medium. As such, matter is a natural resonating cavity of such waves, and in certain geometrical alignments and dimensions, that resonance can conceivably be made to function more efficiently. Desborough's insight is keen, but unfortunately, neither argued nor substantiated.

So inevitably the question occurs: is there such a gridwork of sites upon the Earth? And do temples in fact evidence a geometry in their very construction that would enable them to function as such resonators?

A. The Modern Rise of Earth Grid Theories
1. Ivan Sanderson

One does not have to read too far into the area of Earth Grid theories to find a veritable overgrowth of wild speculations and fanciful ideas. But the actual modern growth of the theory may best be attributed to the scientific research of some Russian chemists, and a Western scientist interested in "anomalous phenomena," the famous Fortean researcher Ivan Sanderson. Sanderson was in fact a professional biologist by academic training and background. But in 1972 he published an article for *Saga* magazine entitled "The Twelve Devil's Graveyards Around the World."

What Sanderson did was to compile a database of "Bermuda Triangle"-like phenomena of ship and plane disappearances worldwide, and, using modern techniques of communications and geophysical tabulation, plotted these on a map of the globe. These correlated to 12 areas on the globe where magnetic anomalies "and other energy phenomena were linked to a full spectrum of strange physical phenomena."[4] Plotting these strange occurrences on a map yielded 12 regions where such phenomena occurred with a greater statistical occurrence than usual:

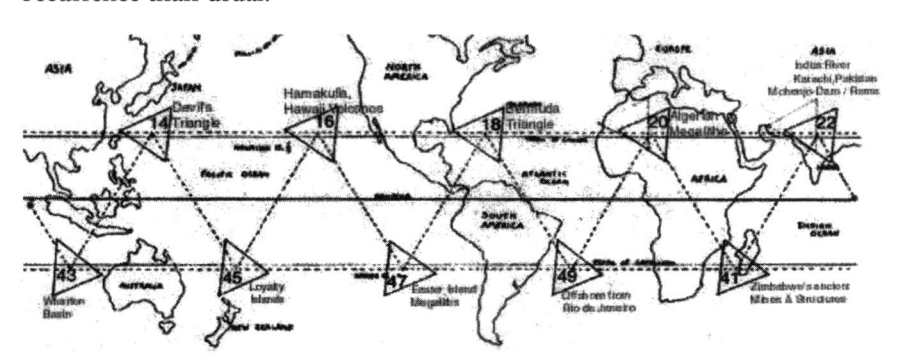

Ivan Sanderson's Map Showing the 12 "Vile Vortices"

---

[4] www.vortexmaps.com/devils-triangle.php

What Sanderson also noted, however, was that some of these areas also included strange megalithic structures. For example, on the map above, in the Pacific Ocean west of South America, one such triangle incorporates Easter Island, with its well-known and very strange stone carvings of gigantic stylized human heads. Another, located in the Algerian desert, also was home to megalithic structures.

2. The Russians Get into the Game

Sanderson's findings spurred a veritable flurry of activity, as other researchers, notably the engineer and mathematician Carl Munck, quickly noticed that virtually all ancient sites were laid out according to a scheme incorporating some sort of "Earth Grid" including the tenets of sacred geometry and ancient units of measurement such as the "megalithic yard." No less than the Soviet Academy of Sciences became involved in this effort, when an article, "Is the Earth a Large Crystal?" appeared in its journal *Khimiya I Zhizn (Chemistry and Life)*, authored by a trio of very unlikely collaborators.

One of them, Nikolai Goncharov, was an historian fascinated by the ancient world and its history. Goncharov plotted on a map of the world all the "centers of earliest human culture."[5] Goncharov then met a construction engineer, Vyacheslav Morozov, and Vaelry Makarov, an electronics researcher. Pooling their resources, the trio, following the lead of some Soviet scientists, suggested the Earth actually began as a crystal "with angular dimension" that only after "millennia of motion and the actions of many forces did the crystal round itself into a ball." Moreover, because of this, hidden beneath the surface of the Earth, the edges of that crystalline structure were buried and possibly still faintly visible.[6]

Later, researchers Bill Becker and Bethe Hagens expanded on the Russian article with their own version. Hagens described her own reaction to the Russian research, and its implications, as follows:

> I found a picture of the world with a funny lattice work on it. It looked as if someone had put one of Buckminster Fuller's domes over the planet, and the design caught my eye. As I read the accompanying article...I learned that was indeed what had been done. Three Russians (an engineer, a historian, and a linguist) had found that the dome-like geometric pattern could be aligned on the Earth in such a way that the struts of the dome mapped out major geologic features (such as

---

5 www.vortexmaps.com/grid-history.php
6 www.vortexmaps.com/grid-history.php, pp. 1–2.

mountain ranges and river systems), and the connecting points for the struts fell at the sites of important ancient civilizations.[7]

Hagens and Becker then expanded on the Russian concept, and produced the following interesting map, which indeed looks like some bizarre creation of Buckminster Fuller:

The Original Russian "Earth Crystal"

Flattening the projection, and adding Becker's and Hagen's research to it, yields their Earth grid map:

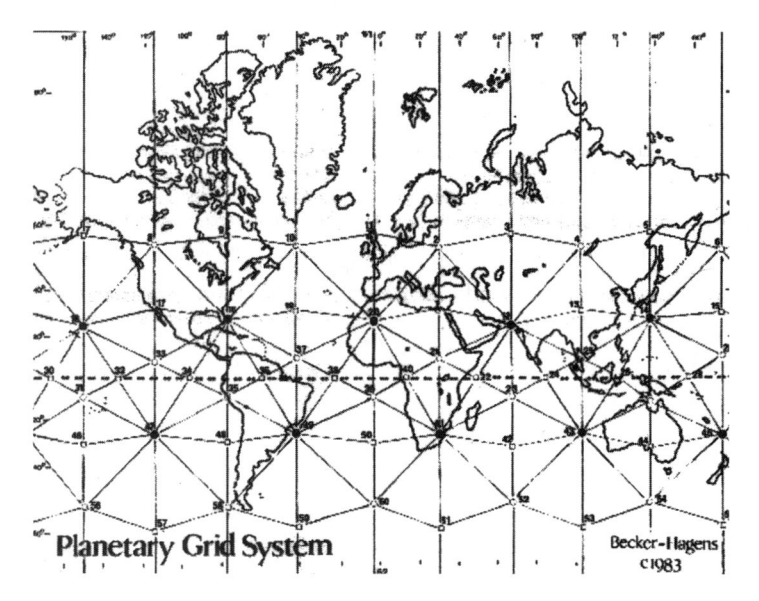

The Becker-Hagens Earth Grid Map of 1983

---

7 Ibid., p. 2.

Looking at this map carefully discloses some rather interesting features. For one thing, the convergence of several lines in the southern Pacific Ocean west of Latin America is, once again, close to Easter Island and its megalithic stone statues. Yet another line runs due north and south through England, close to the celebrated structure of Stonehenge. Yet another node of convergences occurs in Florida, near the well-known Coral Castle, constructed — no one knows exactly how — by Edward Leedskalnin. Finally, yet another node occurs at almost the exact location of the Giza plateau in Egypt. And lest one think this was the fanciful creation of New Age dabblers and cranks, let it be noted that William Becker was a professor of industrial design at the University of Illinois, while Bethe Hagens was a professor of anthropology at the Governors State University of Illinois.[8]

The mention of Buckminster Fuller affords a significant clue into what might be going on with this Earth Grid system. In a series of unusual experiments, Fuller decided to test what sorts of wave patterns might emerge within spheres when subjected to acoustic stress, that is, when subjected to stress of longitudinal waves. Fuller painted the surfaces of balloons, immersed them in water, and then pulsed the water with sound waves of varying frequencies. Fuller discovered that as the spheres vibrated, they set up standing waves on the surfaces of the balloons such that the paint would begin to collect and form lines and grid-like patterns on the surface.[9]

With this one has the conceptual link to the ideas of Kozyrev, Bohm, and others, for if matter is the *result* of the inference pattern of such "longitudinal waves in the medium" (or, as some have called them, "scalar" waves") and as such is therefore a natural resonator of them, then it stands to reason that as a resonator, such waves will be perpetually established within any planetary body, producing such a gridwork. Small wonder then that Tesla, in his scheme for the beaming of wireless *power* to any part of the globe, found it necessary to "grip the earth," for such waves, resonating with the ever-changing dynamism of the planets themselves, would be a virtually inexhaustible supply of energy. And small wonder too, that J.P. Morgan, with his own deep connections to the oil industry and its understanding of energy as a non-renewable closed system of a scarce resource, eventually shut Tesla down.

3. Back to the Nazis

But lest it be forgotten, the first major world power to show an interest in

---

8   www.soulsofdistortion.nl/SOD_chapter7.html, p. 1
9   Similar experiments were conducted by the physician Hans Jenny using a vibrating plate and ordinary sand, with several beautiful structures resulting in a gridlike structure of standing waves. These pictures and the basis of Jenny's study form the subject of a fascinating book, *Cymatics.*

Earth grid theories was not Soviet Russia, but Nazi Germany. As detailed in my previous book *The Philosophers' Stone,* "Himmler's Rasputin," Karl Maria Wiligut, introduced Himmler to the geomantic ideas of Günther Kirchhoff, who believed that there was such a world grid of ancient sites laid out on power points designed to draw energy from the earth itself.[10]

B. Dr. Konstantin Meyl's Paleophysical Interpretation of Ancient Temples as Scalar Resonators

If the conception of matter itself, as a "template" or gridwork of the interference of such longitudinal standing waves in the medium rationalizes, is capable of rationalizing the *placement* of certain ancient temples on the surface of the Earth, what about the incorporation of sacred geometry into their very dimensions?

One scientist who tackled this problem very directly is German physicist and engineer Prof. Dr.-Ing. Konstantin Meyl. Meyl is the author of probably the only comprehensive — and highly mathematical — textbook treatment of the production of such scalar or longitudinal waves in the medium. The title of Meyl's book, all 654 pages of it, is *Scalar Waves: From an Extended Vortex and Field Theory to a Technical, Biological, and Historical Use of Longitudinal Waves.*[11] For the technically and mathematically minded, this book is a thorough introduction to the theory and practice of scalar waves, if one can get through the sometimes shaky English translation.

Meyl is a scientist who very deliberately and consciously sets out to reconstruct an underlying physics basis — what I have called "paleophysics" in my various books on ancient history and texts — from ancient myths, texts, and temples. And his program, as outlined in the final pages of his book, is a total one.

For example, he reproduces the following table as a summary of his particular way of "decoding" the ancient temples and mysteries:

Ancient temple	=	short wave station
Dedicated to one god	=	fixing of the frequency
Supreme god Zeus, Father of all gods	=	range of the short wave / all short wave bands

---

10 Joseph P. Farrell, *The Philosophers' Stone: Alchemy and the Secret Research for Exotic Matter,* pp. 254–255.
11 Villingen-Schwennigen, Germany, 2003.

Priest, Representative of the god	=	amateur radio operator with license to transmit
High priest	=	chief intendant
Pontifex Maximus, "topmost bridge builder"	=	Chairman of the authority and the telegraph offices
Oracle	=	telegraphy receiver
Runes, cuneiform writing	=	telegraphy symbols
Metre, hexameter	=	increase of redundancy
Oracle priest	=	telegraphy interpreter
Tripod	=	reception key, electro-acoustical converter
Looking at intestines, Rite of sacrificing	=	reading off convulsions, electro-optical converter
Temple books	=	news notes
Seer, who looks into the God world	=	amateur radio operator, at telegraphy reception
Homer	=	Ancient radio operator
Godology	=	high-frequency technology
God name	=	RDS, station identification
Members of a family of gods	=	broadcasting studios of a broadcasting company
…		
Earth radiation	=	power supply

Homage of a weekday	=	time restriction of the operation of the station
Zeus forges "thunderbolts"	=	electrostatic blows, when a Temple is oscillating
Ritual act	=	technical provision for transmission and reception
Cella (marrow of temple)	=	tuned cavity
Obelisk	=	antenna rod[12]

While this table of comparisons and decodings might at first glance seem highly implausible if not downright bizarre, Meyl minces no words in the explanatory text accompanying this table:

> It shall be proven that already in antiquity radio engineering based on scalar waves has been used. The proof starts with a thesis.
> The temples in antiquity were all short wave broadcasting stations. And energy from the field served as an energy source, so e.g. the earth radiation in the case of temples of terrestrial gods. In the case of the solar god the radiation of the Sun was used, whereas for the temples, which were dedicated to the planetary gods, the neutrino radiation arriving from the planets served as an energy source.[13]

Meyl's program, in other words, is a total one. In terms of the context of the hypothetical scenario with which we opened this book in chapter one, however, it is also an intriguing one, for given the close association of the ancient bullion brokers with the temples, they, like their modern counterparts, would have need of a vast communications network, one, moreover like their modern counterparts, that would be faster than those communications available to the common masses. So Meyl's thesis, from one standpoint, fits the needs of the situation perfectly.

It is, however, when Meyl turns to a consideration of the sacred geometry in these temples that his thesis takes on its breathtaking significance. In order to buttress his thesis, Meyl reproduces the diagrams of the floor plans, and in some cases, side views and cutaways, of several ancient temples. We do so here in order to illustrate his thesis, and to set his remarks and commentary into

---
12 Prof. Dr.-Ing. Konstantin Meyl, *Scalar Waves,* pp. 608, 610.
13 Ibid., p. 609.

their proper context. Meyl first reproduces the floor plan and front view of the Temple of Zeus in Olympia, showing its use of the Golden Ratio, or Phi ($\phi=0.61818...$):

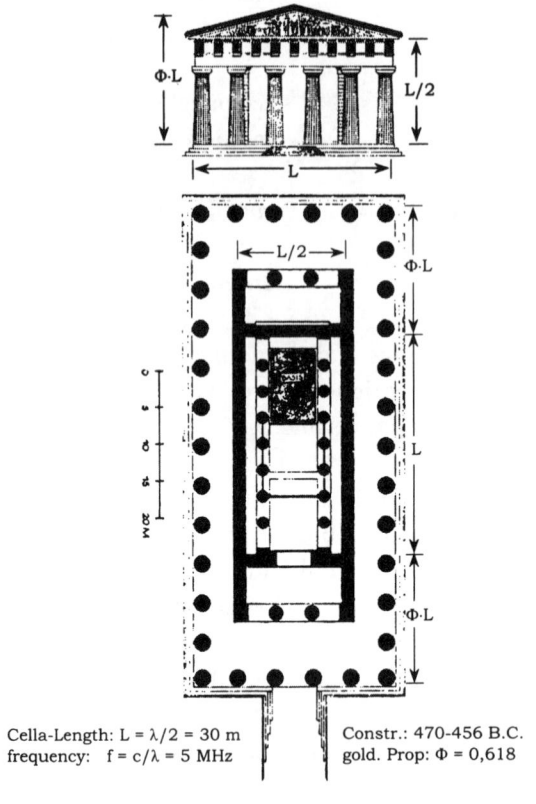

The Temple of Zeus at Olympia, with Meyl's Calculation of its Resonant Frequency[14]

Note Meyl's calculations of the resonant frequency of the Temple on the lower left of his diagram: 5 MHz or 5 Megahertz. On the right, notice how the various dimensions of the Temple reflect the harmonics of the width of the Temple, with the width being designated by L, and the various harmonics of that fundamental being L/2, and ϕ-L, in a kind of stone chamber version of Dewey's organ pipes from chapter two. This analogy is not as haphazard as it might at first be seen, but in order to appreciate why this is so, more of Meyl's diagrams need to be examined.

The next temple Meyl reproduces is a frontal floor plan and side view of the Temple of Athena Alea, ca. 350 B.C.:

---

14 Meyl, *Scalar Waves*, p. 612.

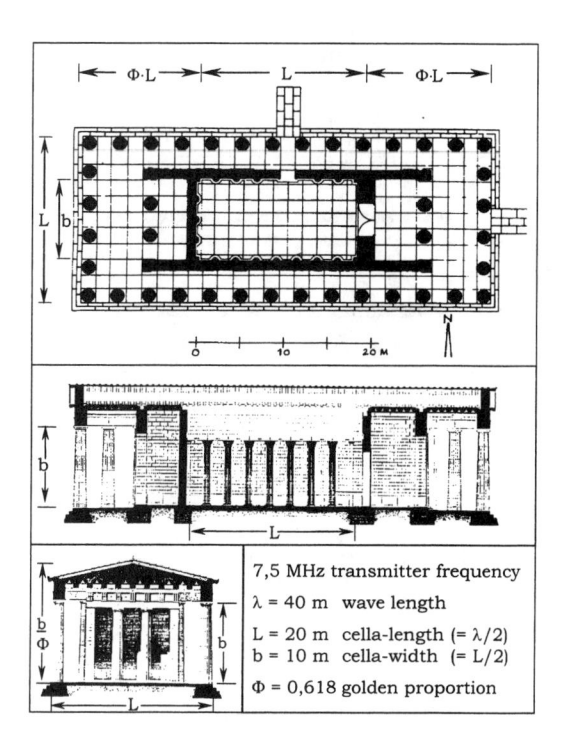

Temple of Athena Alea, 350 B.C.[15]

Note again that the Temple is laid out on redundant harmonic relationships of a fundamental, L, representing the width of the Temple, again with the harmonic ϕ-L being a prominent feature of the structure. Note also Meyl's calculations of the resonant wavelength of the structure being 40 meters, with a "transmitter frequency" of 7.5 Megahertz. He reproduces a similar diagram for the Hera Temple of Selinunt, ca. 460 B.C.:

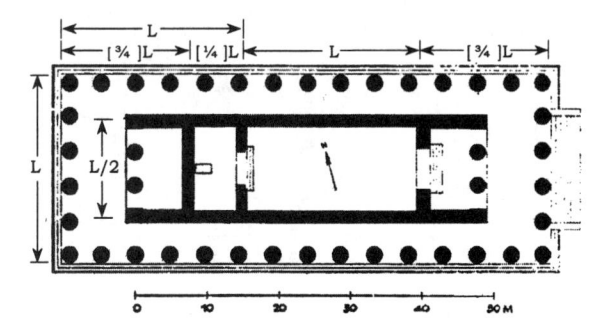

Hera Temple of Selinunt, ca. 460 B.C.[16]

---

15   Meyl, *Scalar Waves*, p. 614.
16   Meyl, *Scalar Waves*, p. 618.

and the Temple of Apollo at Corinth:

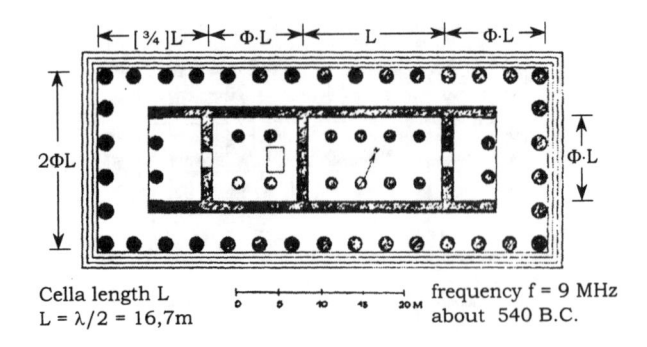

```
Cella length L                    frequency f = 9 MHz
L = λ/2 = 16,7m                   about 540 B.C.
```

Apollo Temple at Corinth[17]

Similar harmonic relationships also held true for later Roman temples, such as the temple of Venus and Roma at Rome, ca. 136 A.D.:

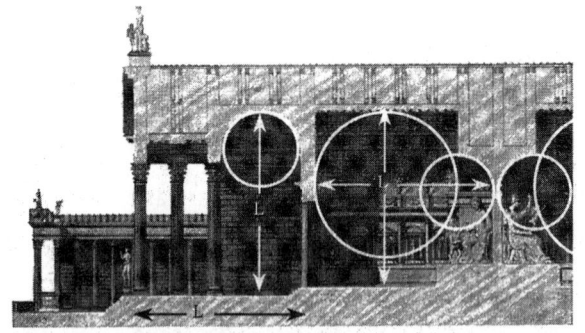

Diameter bigger circle D_1= 22 m (6,8 MHz),
small circle D_2 = 11 m; und L = D_1 + ½ D_2 = 27,5 m (5,5 MHz)

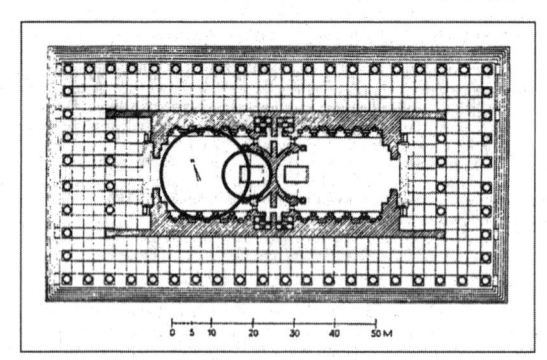

Temple of Venus and Roma, ca. 137 A.D.[18]

---

17 Ibid.
18 Meyl, *Scalar Waves*, p. 620.

And finally, of course, there was the temple of temples, the Roman Pantheon itself:

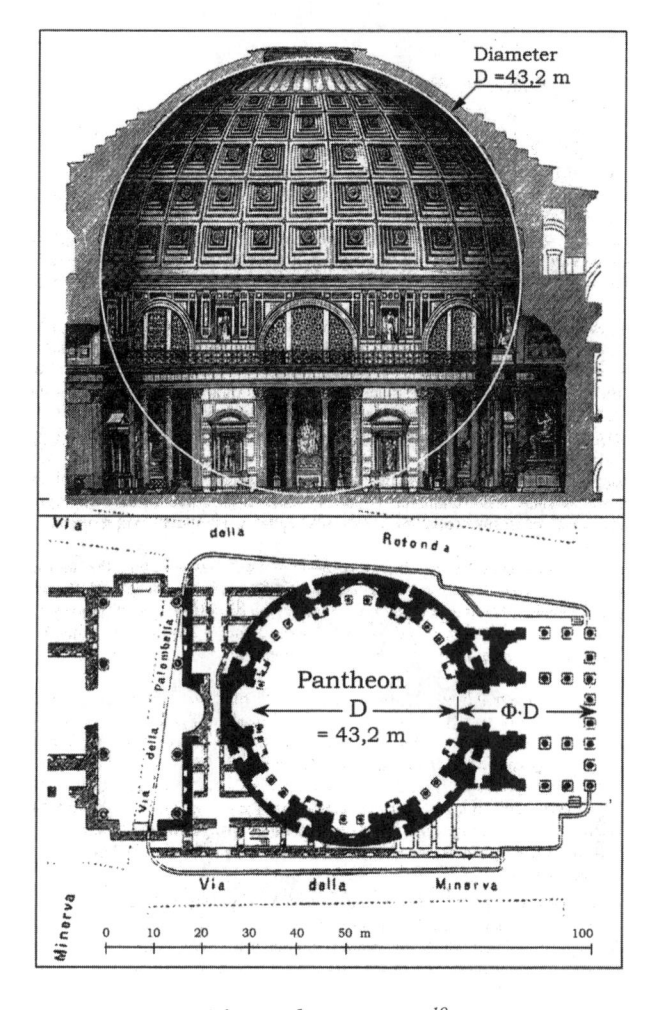

The Pantheon in Rome[19]

Looking carefully at the floor plan of the Pantheon, however, Meyl noticed something that, to his engineer's eye, looked extraordinarily familiar, and a breathtaking confirmation of his thesis that ancient temples may have been deliberately designed as transmitters and receivers of scalar "radio" waves.

Deciding to see if there were other similar floor plans, Meyl went in search of them, and soon found them, one in the floor plan of a temple in the palace of the Emperor Diocletian:

---

19 Meyl, *Scalar Waves*, p. 624.

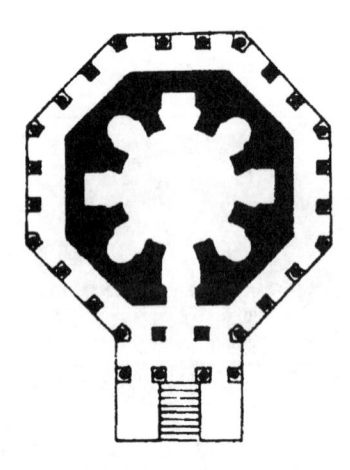

Floor Plan of the Temple in the Palace of Emperor Diocletian[20]

and yet another example in the floor plan of the Temple of Minerva Medica in Rome, built ca. 320 A.D.:

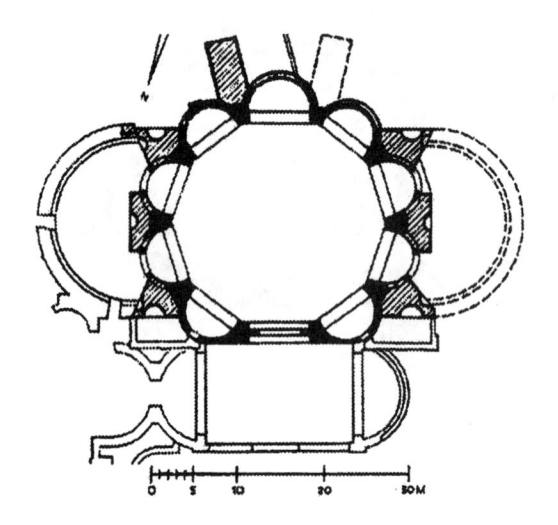

Temple of Minerva Medica in Rome, ca. 320 A.D.[21]

What precisely was so amazing about these floor plans of late Roman temples?

It was their suspicious resemblance to a modern device — crucial in the operation of radios and radars — called a magnetron:

---

20   Meyl, *Scalar Waves*, p. 622.
21   Ibid.

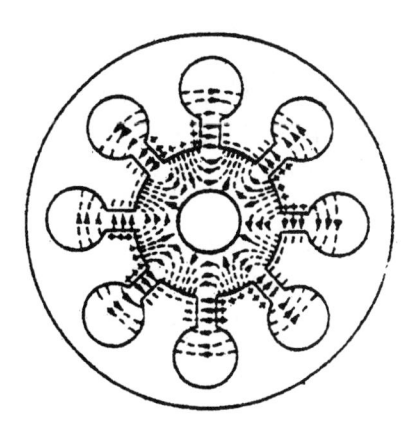

A Modern Magnetron[22]

Meyl's commentary accompanying these diagrams is now worth citing, for it shows how strong the case actually is that some of these ancient temples, a good many in fact, may have been covertly designed for the precise purposes of long-range communication, using some very advanced notions of physics:

> Let's to some extent proceed from the knowledge of textbook physics currently present in high frequency engineering and give a well trained engineer the following task, which he should solve systematically and like an engineer. *He should build a transmitter with maximum range at minimum transmitting power,* thus a classic of optimization. *Doing so, the material expenditure doesn't play a role!*
>
> After mature deliberation the engineer will hit upon it that only one solution exists. He decides on a telegraphy transmitter at the long wave end of the short wave band, at f=3 MHz, which corresponds to a wavelength of λ=100m. There less than 1 Watt transmitting power is enough for radio communication once around the earth....
>
> And he optimizes further. Next the engineer remembers that at high frequencies, e.g. for microwave radiators, not cables but waveguides are used, since these make possible a considerable better degree of effectiveness. In the case of the waveguide the stray fields are reduced by alignment and concentration of the fields in the inside of the conductor. In the case of antennas however the fields scatter to the outside and cause considerable stray losses. He draws the conclusion that his transmitter should be built as a tuned cavity and not as an antenna!

---

22   Meyl, *Scalar Waves*, p. 622.

> As a result the engineer puts a building without windows in the countryside with the enormous dimensions of 50 m length (=λ/2) and 25 m (=λ/4) respectively 12.5 mm (=λ/8) width. The height mhe calculates according to the Golden Proportion to increase the scalar wave part. *Those approximately are the dimensions of the Cella without windows of Greek temples.*
>
> For the operation of such a transmitter in antiquity apparently the noise power of the cosmic radiation was sufficient, which arrived at the earth starting from the sun and the planets. By increasing the floor space also the collected field energy and the transmitting power could be increased, so that also from the perspective of the power supply the temple with the largest possible wavelength at the same time promised the largest transmitting power, so at least in antiquity.
>
> Our engineer further determines that he will switch the carrier frequency on and off at a predetermined clock pulse. Thus he decides for radiotelegraphy. The advantage of this technique is a maximum of the increase of the reception range. For that the signals at the transmitter have to be coded and at the receiver again deciphered. By means of the encryption of the contents these are accessible only to the "insiders," who know the code; *prerequisite for the emerging of hermeticism and eventually a question of power!*[23]

This leads Meyl to his next bit of evidence, or rather to the next stage of his decoding of the ancient data pointing to the existence of a genuine technology at work: ancient texts.

Meyl, having demonstrated the peculiar transmitter-receiver properties of many ancient temples, then uses the existence of this technology to decipher a baffling mystery concerning ancient texts:

> Direct evidence is present as well. It can be found in ancient texts. But it is questionable if historical texts concerning ancient radio engineering have been translated correctly. The talk is about oracles, mystery cult and earth prophecy if the receiver is meant. The predominantly technically uneducated historians attest (that) the Romans (possessed) a defective sense of time, because their couriers surely could not cover the long ways across the Roman empire so fast at all, if the read in the Latin texts: "They sent by courier to the emperor in Rome, and got for answer…". The answer of the emperor namely already arrived at the squad at the latest in the following night. The correct translation

---

23 Meyl, *Scalar Waves,* emphasis added, p. 613.

should read: "they cabled" or "they broadcasted to the emperor in Rome and got for answer..."

Such a big empire as the Roman Empire actually only could be reigned by means of an efficient communication. Cicero coined the word: "We have conquered the peoples of the earth owing to our broadcasting technology..."! The term broadcasting technology from ignorance is translated with piety. If engineers however rework the incorrect translations, then one will discover that numerous texts tell of the broadcasting technology, that thus correspondingly much direct evidence exists concerning the practical use of this technology.

For the Roman military transmitters, which formed the backbone of the administration of the empire, the reading off of the information from observations of nature like the bird flight or from felt signals of a geomanter was too unreliable. They read off the information from the rhythm of the convulsions of the intestines of freshly slaughtered animals. In the case of the dead animals on the altar every extrinsic influence was excluded. But the enormous need of slaughter cattle was a disadvantage. Who wanted to have information first of all had to bring along an animal, which then was "sacrificed" to the god, or better say, which was abused as a receiver of a particular transmitter. Thereby the innards served as a biosensor and as a receiver for the news.[24]

In other words, an extraordinarily sensitive membrane had to be found to be able to sense and record the very subtle pulses of such transmissions and telegraphy, for they were by the nature of the case, very weak.

Whatever one makes of these speculations of Meyl, he is, however, correct in his basic premise that the design of ancient temples is too coincidentally peculiar to the engineering requirements of transmitting weak radio signals. Nor should the *material* used to construct such temples be overlooked, for many of them were constructed from limestone, granite, and other crystal-bearing rock, and the ability of crystals in receiving radio signals is common knowledge.

But why does he invoke the scalar component in his argument that such temples were, in fact, tuned resonant cavities? The key lies with what seems to be the most absurd component of his reconstruction, namely, the use of the membranes of an animal's intestines to record the pulses, for the pulses being received are *not* ordinary Hertzian "jump rope" waves, but the "yardstick" waves of *longitudinal or scalar waves:*

---

24 Meyl, *Scalar Waves,* p. 615.

The argumentation has to be made on mathematical-physical foundation. The prerequisite for that are the 29 chapters of before.[25] The following points could be demonstrated and derived:

1. The wave equation (inhomogeneous Laplace equation) describes the sum of two wave parts, where
2. every antenna emits both parts.
3. The transverse part, known as electromagnetic wave (Hertzian wave)
4. and the longitudinal part (Tesla radiation) termed scalar wave by the discoverer, better known as antenna noise.
5. The wave equation mathematically describes the connection of both wave parts in general and the conversion of one part into the other in particular, thus
6. the rolling up and unrolling of waves in field vortices (measurable as noise).
7. The transition takes place proportionally to the Golden Proportion, as resulted from the derivation (chapter 29.7–29.9).

With the last point the electrotechnical problem becomes a geometrical problem, if it concerns the use of scalar waves. The geometry of the antenna is crucial. Thereby the Golden Proportion provides the necessary direction for construction.

That justifies the assumption that the buildings in antiquity, which were built according to the Golden Proportion, were technical facilities for scalar waves. Maybe the builders had specifications that had physical reasons and could mathematically be proven.

At this place here result completely new aspects for judging and interpreting buildings especially from antiquity through the derivation of the Golden Proportion from the fundamental field equation. If we have understood their way of functioning, then we will be able to learn much from that for our own future and for the future construction of scalar wave devices.

As a further prerequisite for the ancient broadcasting technology enough field energy should be at disposal. We proceed from the assumption that

1 the earth magnetism and the cosmic neutrino radiation are tightly banding together by the processes in the earth's core,

---

25   In other words, the entirety of his book which preceded this concluding section!

2. the earth magnetism in antiquity verifiably was (approximately a) thousandfold stronger than today (proven by gauging of pieces of broken pot),
3. as a consequence the neutrino radiation in antiquity as well must have been (a) thousandfold strong and
4. the cosmic neutrino radiation has served the transmitting plants of antiquity as an energy carrier.

Any thought is absurd to reject the technical function of a temple only because it today can't be reproduced anymore. The artistic and aesthetical viewpoints, which are put into the foreground by art historians because of ignorance about the true function, rather are secondary.[26]

And with those rather breathtaking statements, Meyl's case for the ancient temple as being a broadcast technology based on longitudinal pulses or scalar waves is concluded.

However, in view of my own speculative analysis of the Earth grid itself as being the *result* of its natural properties of being a resonator of such standing longitudinal waves, one is inevitably led to the conclusion that *both* the use of "sacred geometry" in the construction of such temples, *and* the placement of some of these temples over precise nodal points on that gridwork was for the explicit and precise purpose of making these structures as efficient resonating cavities as possible. When one adds to this the fact that the resulting template or interference pattern of such longitudinal waves, produced by all such resonators, such as the planets, vary with time according to their positions relative to each other, and to the cosmic backdrop of the galaxy, one again finds a physics reason why so many of these very same temples were oriented to astronomical points or events.

To put it succinctly, it is beginning to look an awfully lot like that Very High and Very Ancient Civilization, having blown itself apart by means of those very same technologies, moved quickly and effectively to preserve as much of that technology as it could in the establishment and propagation of the ancient temples and mystery schools, and moved equally quickly to ensure that its legacy civilizations knew and understood the necessity — if there was to be any kind of civilization at all — for there to be a medium of exchange based upon the creative and productive output of the physical medium itself, and of its derived and differentiated creatures: man. Little wonder, too, that those who sought to reconstruct that technology in order to gain mastery and

---

26   Meyl, *Scalar Waves*, p. 611.

hegemony over their fellow man quickly allied themselves with those temples and mystery schools, and begin to pervert that physics in a financial and false alchemy that created the exact opposite: "negative information" in the form of money-as-debt interest, for by means of that closed system of finance, they purposed to reconstruct that open system of physics for themselves, to ensure their power and hegemony.

For with living organisms — as Dr. Meyl himself alluded to — and more importantly, mankind, one arrives at the final component and resonant cavity of this form of energy: human DNA, and its own remarkable embodiment of the very same "Sacred Geometries."

Eight

TEMPLATES, GENOMES, AND BANKSTERS
OR, WHY DO THEY ALL SEEM TO MARRY COUSINS AND END UP WITH COLOSSALLY STUPID KIDS?

❖

> *"Inbreeding in European royal families has declined slightly in relation to the past. This is likely due to clear scientific evidence of genetic degeneration."*
> —"Inbreeding," Wikipedia[1]

Both in ancient and in modern times we have discerned the persistent outline of an association of an international class of banksters with those engaged in advanced scientific research, and we have also seen that this class has endeavored to monopolize not only the false alchemical power of the issuance of debt as a facsimile of money, but also to monopolize the genuine alchemical power of the physics of the transmutative physical medium itself. But what about the period *between* these two widely separated historical poles? Did that class continue to exist? And if so, what were they doing? Is there evidence to suggest that it continued as a class in more or less uninterrupted continuity from ancient times? And is there evidence to suggest that it continued its close association and monitoring of scientific achievement?

Moreover, it was seen that some allege that the Rothschild family secretly traces their family origins back to the Sumerian conqueror Nimrod. Is there any truth to, or for that matter, any broad corroboration of, these allegations?

A. Ancient Rome

To answer these questions, a closer look at ancient Rome is in order.

---
1 "Inbreeding," Wikipedia, en.wikipedia.org/wiki/Inbreeding, p. 4.

Professor Tenney Frank was a professor at Johns Hopkins University at the turn of the last century, and was the author of a well-known treatise, *An Economic History of Rome*. This work became such a standard in the field that it became the basis for entries in the *Cambridge Ancient History* and the *Oxford History of Rome*.

In the course of his researches, Professor Frank discovered an interesting thing about the population of ancient Rome during the period between the Republic and the final emergence of the empire, and that is that the population was not Roman or Latin at all, but — in a word and without much exaggeration — Babylonian. In an article written for the *American Historical Review*, Frank begins by noting how the problem of the ancient Roman race came to his attention. Visiting the ancient tombs of Rome, Frank observes that the historian will notice a curious and even peculiar thing:

> …he finds prenomen and nomen promising enough, but the cognomina *all* seem awry. L. Lucretius *Pamphilus*, A. Aemilius *Alexa*, M. Clodius *Philostoasgas* do not smack of freshman Latin. And he will not readily find in the Roman writers now extant an answer to the questions that these inscriptions invariably raise. **Do these names imply that the Roman stock was completely changed after Cicero's day, and was the satirist (Juvenal) recording a fact when he wailed that the Tiber had captured the waters of the Syrian Orantes?** If so, are these foreigners ordinary immigrants, or did Rome become a nation of ex-slaves and their offspring?[2]

Frank's tentative answers are even more unsettling than the original questions he asks. Studying almost 14,000 names, Professor Frank discovered that between 300 B.C. and 300 A.D., the population of Italian Rome underwent a drastic change in ethnicity, such that by the end of the period, the vast majority of the Italian Roman population had Greek surnames, and not Latin at all:

> For reasons which will presently appear I have accepted the Greek cognomen as a true indication of recent foreign extraction, and since the citizens of native stock did not as a rule unite in marriage with (freemen), a Greek cognomen in a child or one parent is sufficient of status (i.e., was foreign).[3]

Professor Frank quickly dispatches the idea that this was simply due to a "popularity" craze of certain names:

---

2 Tenney Frank, *The American Historical Review*, Vol. 21, July 21, 1916, p. 689, italicized emphasis in the original, boldface emphasis added.
3 Tenney Frank, *The American Historical Review*, Vol. 21, July 21, 1916, p. 681.

On the other hand, the question has been raised whether a man with a Greek cognomen must invariably be of foreign stock. Could it not be that Greek names became so popular that, like Biblical and classical names today, they were accepted by the Romans of native stock? In the last days of the empire this may have been the case, but the inscriptions *prove* that the Greek cognomen was not in good repute. I have tested this matter by classifying all the instances in the 13,900 inscriptions where the names of both father and son appear. Form this it appears that fathers with Greek names are very prone to give Latin names to their children, whereas *the reverse is not true.*[4]

Thus, the conclusion for Professor Frank was rather obvious:

Clearly the Greek name was considered as a sign of dubious origin among the Roman plebeians, and the freedman family that rose to any social ambitions made short shrift of it. For these reasons, therefore, I consider that the presence of a Greek name in the immediately family is good evidence that the subject of the inscription is of servile or foreign stock. **The conclusion of our pros and cons must be that nearly *ninety per cent* of the Roman-born folk represented in the abovementioned sepulchral inscriptions are of foreign extraction.**[5]

But this posed a rather large problem, and to see what it is and how Professor Frank dealt with it, we need to recall a bit of obvious history.

The problem is the mere presence of a Greek surname on tomb inscriptions does not necessarily mean the occupant is *Greek*. The simple reason for this is that after Alexander the Great's conquest of Mesopotamia, Persia, and Egypt, the Greek language spread far and wide, and became something of an "international" language, much as English or French is today. Thus, a Greek surname on a tomb does not necessarily imply that all these Roman tombs had Greek occupants. But this raises a far larger and more significant question. If they were not necessarily Greek, then what were they? What was the predominant ethnic stock they represented? Frank minces no words:

Who are these Romans of the new type and whence do they come? How many are immigrants, and how many are of servile extraction? *Of what race are they?*[6]

---
4 Ibid., pp. 692–693.
5 Ibid., p,. 693, italicized emphasis original, boldface emphasis added.
6 Tenney Frank, *The American Historical Review*, Vol. 21, July 21, 1916, p. 693.

Frank's method of dealing with this question is to take Rome's classical authors and satirists at their word, and from this, an important and very significant fact emerges.

Noting that "most of the sociological and political data of the empire are provided by satirists," Frank then reasons himself to a rather astonishing conclusion:

> When Tacitus informs us that in Nero's day a great many of Rome's senators and knights were descendants of slaves and that the native stock had dwindled to surprisingly small proportions, we are not sure whether we are not to take it as an exaggerated thrust by an indignant Roman of the old stock…. To discover some new light upon these fundamental questions of Roman history, I have tried to gather such fragmentary data as the corpus of inscriptions might afford. The evidence is never decisive in its purport, and it is always, by the very nature of the material, partial in its scope, but at any rate it may help us to interpret our literary sources to some extent. *It has at least convinced me that Juvenal and Tacitus were not exaggerating.* It is probably that when these men wrote a very small percentage of the free plebeians on the streets of Rome could prove inmixed Italian descent. *By far the larger part, perhaps ninety percent, had Oriental blood in their veins.*[7]

One has only to read a bit between the lines to see what Professor Frank is implying, for Juvenal, let it be recalled, had complained of the Syrian "Orantes" river "flowing into the Tiber," a metaphor for people of Chaldean — i.e., *Babylonian* — extraction having "flowed" into the bloodlines of the ancient Roman stock: "These dregs call themselves Greeks," he complains, "but how small a portion is from Greece; the River Orantes has long flowed into the Tiber."[8]

The basic historical outlines are now clear, for as Roman conquests spread into the eastern Mediterranean and eventually conquered the old Seleucid Empire — i.e., the portion of Alexander's empire based in Mesopotamia with its capital at Babylon — many of these peoples made their way back to the Italian peninsula as slaves, and, following the relatively lenient Roman custom of manumission of slaves upon the death of their owner, these later became freemen and the backbone of the Roman economy in the very lap of the Empire itself.

---

7 Ibid., pp. 689–690, emphasis added.
8 Juvenal, *Satires,* III: 62.

1. The Change in Roman Racial Stock and Imperial Policy

This brings us to the economics and finance part of the story, the part that is for our purposes, the most significant part:

> There are other questions that enter into the problem of change of race at Rome, for the solution of which it is even more difficult to obtain statistics. For instance, one asks, without hope of a sufficient answer, why the native stock did not better hold its own. Yet there are at hand not a few reasons. We know for instance that when Italy had been devastated by Hannibal and a large part of its population put to the sword, immense bodies of slaves were brought up in the East to fill the void; and that during the second century B.C., when the plantation system with its slave service was coming into vogue, the natives were pushed out of the small farms and many disappeared to the province of the ever-expanding empire. Thus, during the thirty years before Tiberius Gracchus, the census statistics show no increase. During the first century B.C., the importation of captives and slaves continued, while the free-born citizens were being wasted in the social, Sullan, and civil wars. Augustus affirms that he had had a half a million citizens under arms, one-eighth of Rome's citizens, and at that the most vigorous part. During the early empire, twenty to thirty legions, drawn of course from the best free stock, spent their twenty years of vigor in garrison duty while the slaves, exempt from such services, lived at home and increased in numbers. In other words, the native stock was supported by less than a normal birthrate, whereas the stock of foreign extraction had not only a fairly normal birthrate, but a liberal quota of manumissions to its advantage.[9]

The result of this combination of bad policy, wars, and heavy reliance on slaves, was that the original Roman race — at least on the Italian peninsula — "went under."[10]

However, there were two significant things that this importation of Mesopotamian slaves accomplished. First, these slaves brought with them, of course, their culture and religion. Secondly, they brought with them their "Babylonian" business and banking practices. Thus, even when this population were slaves, the great majority of the normal practical day-to-day commerce of Rome — farming, construction, instruction, and so on — was conducted by slaves. And when freed, not only did their religion and culture penetrate Roman

---

9 Tenney Frank, *The American Historical Review*, Vol. 21, July 21, 1916, p. 703.
10 Ibid., p. 704.

society to a significant degree, but this group gradually penetrated the highest reaches of the Roman imperial government itself.[11] The result, in short, was that the Mesopotamian and Syrian merchants effectively colonized Rome's provinces bordering the Mediterranean Sea, Roman banking was all but monopolized in their hands as the influence of Mesopotamian mystery cults extended throughout the Empire, and the activities of the temple continued to be associated with commerce. In a certain limited sense, then, the Roman Empire may be viewed as but the latest imperial front for that ancient class of "bullion brokers," and this change of the ethnic stock in the heart of the Empire, the Italian peninsula itself, finally, explains why the bullion exchange policies of Rome vis-à-vis the Far East appear to be the product of deliberate manipulation.

2. The Next Stage: Venice and Banking

A quick glance at the subsequent history of this class will show the connections to modern time, for the next stage in this connective history was provided by Attila the Hun who, ravaging the Italian peninsula and even Rome itself, forced many of these mercantile families to flee northward to the protective lagoons and marshes of what would later become the Middle Ages' center of commerce and banking in western Europe: Venice. In the very early ninth century, Venice was formally recognized as a part of the Eastern Roman Empire, and thus began its rise as a financial power, for it was granted special trading and tax exemptions throughout the Empire.

But the pattern, a mercantile and banking class operating behind special imperial privileges, remained the same. As noted in chapter four, however, the Fourth Crusade eventually captured Constantinople, and with that, the Empire's monopoly on the right to coin gold was broken, and western European monarchs began to coin such money. But it was Venice that had funded the mercenary army of French knights that captured Constantinople, and it was the Venetian Doge, Enrico Dandolo, that imposed a puppet Latin government on the ancient Eastern Empire. By this point, the great Venetian oligarchical families and their family *fondi* or fortunes (literally, "funds"), were in place: the Cornaro, the Dandolo, the Contarini, Morosini, Sorzi and Tron fortunes.

3. On to Amsterdam, London, the Reformation, and the Wars of Religion

Researcher Webster Tarpley, in three short paragraphs, outlines the connection of Venetian banking and politics in subsequent European history

---

11 See the *Cambridge Ancient History,* Vol. X, p, 727.

during the Reformation and Counter-Reformation, and the period up to the English "Glorious Revolution":

> Why are the British liberal imperialists called the Venetian Party? Well, for one thing, they call themselves the Venetian Party. The future prime minister Benjamin Disraeli will write in his novel *Conningsby* that the Whig aristocrats of 1688 wanted "to establish in England a high aristocratic republic on the model of Venice, making the kings into doges, and with a 'Venetian constitution.'"
>
> During the War of the League of Cambria of 1509–17, an alliance of virtually every power in Europe threatened to wipe out the Venetian oligarchy. The Venetians knew that France or Spain could crush them like so many flies. The Venetians responded by launching the Protestant Reformation with three proto-stooges — Luther, Calvin, and Henry VIII. At the same time, (Cardinal) Contarini and his Jesuits made Aristotle a central component of the Catholic Counter-Reformation and the Council of Trent, and put Dante and Piccolomini on the Index of Prohibited Books. The result was a century and a half of wars of religion, and a "little dark age," culminating in the Great Crisis of the seventeenth century.
>
> Veince was a cancer consciously planning its own metastasis. From their lagoon, the Venetians chose a swamp and an island facing the North Atlantic: Holland and the British Isles. Here the hegemonic Giovanni party would relocate their family fortunes, their *fondi*, and their characteristic epistemology. France was also colonized, but the main bets were placed further north. First, (Cardinal) Contarini's relative and neighbor Francesco Zorzi was sent to serve as a sex advisor to Henry VIII, whose raging libido would be the key to Venetian hopes. Zorzi brought Rosicrucian mysticism and Freemasonry to a land that Venetian bankers had been looting for centuries....[12]

Reading between the lines, one notes again the familiar and shadowy pattern of the ancient bullion brokers' alliance with the temple, in this case, the temple of Renaissance Catholicism and nascent Protestantism, for one and the same financial power is behind both, and manipulating a conflict for its own profit. More importantly, one notes another sinister and familiar pattern, that of

---

[12] Webster Tarpley, "Lord Palmerston's Multicultural Human Zoo," www.mail-archive.com/ctrl@listserv.aol.com/msg17357.html, p. 4. See also Tarpley, "Venice's War Against Western Civilization," www.schillerinstitute.org/fid_91-96/952_venice.html, pp. 5–7, for a more detailed discussion, particularly of the role of Gasparo Cardinal Contarini in these machinations.

David Rockefeller's "republic" of a global elite of intellectuals and bankers at the apex of a pyramid of power.

B. The Myth of the Rothschild Descent from Nimrod: A Second Look

All this provides an interesting context in which to view the allegations that the Rothschild dynasty secretly traces its family origins back to Nimrod, the conqueror of the Old Testament closely associated with the Tower of Babel incident.[13] The careful reader will have observed throughout this and preceding chapters that there are at least *three different types of historical continuity operating:*

1) The alleged continuity of a bloodline or family, as is alleged for the Rothschilds and their secret family descent from Nimrod, and hence, from the Babylonian "bullion brokers";
2) The continuity of *methods:* alliance with and operation through the temple, cultivation of science and suppression of certain scientific and technological advancement, the promotion of the facsimile of money as circulating privately-created debt, and the policy of such families to intermarry within their own class and, in some cases, their own bloodlines; and,
3) The continuity of *class, culture, family fortunes,* and even — in the obvious descent of Venetian banking from the Oriental stock of slaves imported to Rome — of *ethnic stock.*

Viewing this list, it is evident that so far as the Rothschild Nimrod allegation is concerned, there is a broad corroboration of the pattern as is evident from points 2 and 3. But is there anything more specific that would tend to corroborate point 1, or at least, suggest that it might be true?

Indeed there is.

The famous register of nobility in Britain, *Burkes' Peerage,* records numerous instances of members of the European Rothschild clan marrying each other.

---

13 For this connection, see my *Giza Death Star Destroyed* (2006), pp. 77–78, and *The Cosmic War* (2007), pp. 311–318. See also David Rohl, *The Lost Testament* (2002), pp. 73–76. Rohl clearly identifies the Mesopotamian god Ninurta with Nimrod, based on various considerations. In doing so, he clearly highlights the massive chronological problems that I alluded to in the Preface to *The Cosmic War,* for the archaeological record, as Rohl notes, would *not* support a very ancient dating for the event, such as I implied in that book. Contrariwise, as I also noted, if one takes the ancient texts themselves seriously, and reads them in the backdrop of the Exploded Planet Hypothesis, then such extreme ancientness *is* indicated. I believe the way out of this impasse is once again to recognize that the ancient texts and myths of Sumeria and other cultures may have been deliberately contrived to function on a *multitude* of levels — including chronological ones — simultaneously.

For example, it records that Evelina de Rothschild, daughter of Lionel Nathan Baron de Rothschild and Charlotte de Rothschild, married Ferdinand James Anselm Rothschild, who was the son of Anselm Salomon Rothschild and Charlotte Rothschild.[14] Salomon Albert Anselm Rothschild was son of Anselm Salomon and Charlotte Rothschild, and married Bettina Caroline de Rothschild, daughter of Mayer Alphonse de Rothschild and Leonora de Rothschild.[15]

But interestingly enough, in the midst of this consanguineous interbreeding warren, Alphonse Mayer Rothschild and Clarice Sebag-Montefiore had a son, born in 1922, to whom they gave the peculiarly Jewish-Christian-"Babylonian" name of Albert Anselm Salomon Nimrod Rothschild.[16] The child died only 16 years later, in 1938.

Of course, one male heir in all of the proliferating Rothschild warren is hardly conclusive, but it is suggestive that the name had *some* significance for the clan known only to itself. But there is something else that suggests that the allegation must be taken seriously. Among the banking clan's vast financial network, there is one financial group that raises the eyebrows, and this is the "Rothschild Nemrod Diversified Holdings" group.[17] The name "Nemrod" is of course yet another phoneticization of the name "Nimrod," since the biblical character's name, in ancient Hebrew, was written without vowels as simply NMRD. It can thus be phoneticized as Nimrod, Nimrud, Nemrod, Nemrud, and so on. The use of the biblical conqueror's name for a mutual fund investment group thus connotes aggressiveness, risk-taking, and an intention to dominate and conquer (by financial means, of course).

By why the evident obsession of such families — witness the Rothschild dynasty — with marrying distant relatives and members of their own clans and classes? The answer to that question requires a closer look at human DNA, at its own remarkable connections to sacred geometries and the alchemical physics of the medium, and at the family tree of Nimrod himself.

C. Human DNA and the Hermetic Code
1. The I Ching

Ancient esoteric doctrine held that mankind was a "microcosm," a "little universe" who mirrored in his very constitution — i.e. his size, shape, makeup, and most importantly, in his unique combination of a corporeal component (a body) and a spiritual, or if one prefers, "hyper-dimensional," component

---

14 www.thepeerage.com/p19533.htm, p. 1.
15 Ibid.
16 Ibid., p. 2.
17 www.trustnetoffshore.com/Factsheets/Factsheet.aspx?fundCode=R1F58 &univ=DC.

(his soul and personhood[18]) — the larger universe.[19] British researcher Michael Hayes, taking this doctrine seriously, decided to examine it more closely, and as a result, found a breathtaking connection between the code of sacred geometry and the fascination with certain numbers, and human DNA.

> In order to give myself a kind of visual aid, an image of the code in action, I had drawn up a diagram incorporating the key numbers of the biochemical components involved in the process. These were 4, 3, 64, and 22. That is, there are four kinds of chemical bases. It takes three of them to make what is known as a triplet codon, an amino acid template, of which there are exactly sixty-four variations. Each of these codons correspond to one or another of twenty-two more complex components, namely, the twenty amino acids and the two coded instructions for starting and stopping the process of synthesis. In my diagram, the number 64, the number of triplet-codon combinations (4 x 4 x 4), was represented by a square grid, eight divisions across and eight down, like a chessboard.[20]

Once he had done this, however, Hayes noticed a very odd and striking connection to one of esotericism's oldest systems of divination — the Chinese I Ching: "I realized, in fact, that the whole diagram echoed the format of the famous Chinese work known as the I Ching (Yi King), whose sixty-four basic texts are each identified with a six-line symbol called a hexagram."[21]

But how does Hayes' table and its numbers of 4, 3, 64, and 22, resemble the I Ching in particular, and the Hermetic Code of sacred geometry in general? Hayes explains:

> The I Ching….was intended for use as an oracle: you pose a question, toss three coins, and note the way they fall. A preponderance of heads gives an unbroken line — "yang," positive; tails a broken line, "yin," negative….

The results are written down as a solid or dashed line.

---

18 "Soul and personhood," this phrasing is not redundant, but is meant to express the author's opinion that "soul" and "person" are two very distinct things. How they came to be equated in the popular mind of many Westerners would require a tome in its own right.
19 For a brief survey of this doctrine in the context of other paleophysical themes, see my *Giza Death Star Destroyed,* chapter 3.
20 Michael Hayes, *The Hermetic Code in DNA: The Sacred Principles in the Ordering of the Universe,* pp. 13, 15.
21 Ibid., p. 15.

> Repeat the action six times and you have called up one of the hexagrams. The accompanying text supplies your answer.[22]

In other words, repeating the action six times and writing down each result — a broken or solid line — one above another will produce a picture of solid and broken lines (of which there are only 64 possible results). The pictogram or hexagram is then looked up in a book which has a text explaining the divinatory properties of each hexagram.

But where do the other numbers — 4, 3, and 22 — fit into this pattern?

> Let's begin with the number 4, the number of fundamental chemical bases in the genetic code (adenine, thymine, guanine, and cytosine) upon which the whole process of amino-acid synthesis depends. The I Ching, I discovered, embodies exactly the same principle. The sixty-four hexagrams are actually constructed from four, basic, two-line symbols known as the Hsiang. These in turn were derived from the two fundamental lines, one broken and one unbroken, known respectively as yin and yang.
>
> Next, the number 3. The genetic code, as was evident, obeyed the law of three forces, which is why only triplet codons are evident in the process of creation. The three forces are inititially represented in the Book of Changes by the two original yin (negative, female), yang (positive, male) and neutral, the third, invisible, or "mystical" ingredient: the tao....[23]

Hayes recalls his reaction to this discovery:

> By this time, having recognized so many similarities between the I Ching and the genetic code, I was convinced that I was on to something of profound importance, and my emotional state reflected this: I was highly charged. No way, I thought, could the identical features of these two apparently disparate systems be the product of mere coincidence, for they were not only identical in structure, it seemed that they each had a common purpose, which was to facilitate the process of evolution.[24]

So far, so good.

But what about the number 22? Was there some correspondence between the

---

22 Ibid.
23 Hayes, *The Hermetic Code in DNA,* pp. 15–16.
24 Ibid., p. 16.

I Ching and DNA involving this number? Hayes puts the problem this way:

> Now these hexagrams, as I said earlier, just like the biochemical hexagrams of the genetic code, each consist of two trigrams, two three-line symbols, one above, one below. The trigrams, eight in number, were derived from the four Hsiang, by successively placing over each of them the two original broken and unbroken lines. When these same two lines are placed over the eight trigrams, the result is sixteen figures of four lines. Repeat the process once again and you get thirty-two figures of five lines, and a final similar movement produces the sixty-four hexagrams.
>
> Unlike the four- and five-line figures, the eight trigrams, known as the kua, are given particular prominence in the system. I mused over these for a long time, juggling with their numbers. Eight threes. Three eights. Twenty-four. I needed twenty-two. Close, but not close enough. Certainly the number 8 was an integral part of the overall symmetry, being the square root of that magical 64; but why did the sum of the trigrams not conform to the twenty-two codon signals of the genetic code? Why twenty-four? Why eight?[25]

The answer came when Hayes recalled that the numbers 4, 3, 64, 8 and 22 were significant components to the Hermetic Code and sacred geometries of the *West,* and more particularly, had a direct connection to *music.*

The number 22 was a key number of the Pythagorean system "principally because of its musical aspect."[26] What it represented was in fact "three octaves of vibrations, or notes, three sets of eight — twenty-four components."[27] But where does 22 come in?

The answer is very simple. If one sits at a keyboard instrument, and begins with the note "C," and goes upward eight notes using only the white keys, one will arrive back at "C" an octave higher, exactly eight notes later. Repeat the process again, and at the sixteenth note above the original, one arrives back at "C," and again a third time, and one arrives back on the twenty-fourth note, which is once again "C." But since two of these "C"s are but repetitions of the original "C," one might think of the three octaves as embodying the number 22, as well as 24.[28]

The implications of this were enormous, and Hayes was quick to perceive what they were:

---

25 Hayes, *The Hermetic Code in DNA,* p. 17.
26 Ibid., p. 18.
27 Ibid.
28 Ibid., p. 19.

From that time onward, the summer of 1984, I spent several years exploring the mazelike annals of history. I automatically assumed that, if the Chinese and the Greeks were "tuned in" to this ancient science... then it was probably that so were some of the other traditions and civilizations.... As it turned out, the evidence was overwhelming. Everywhere I looked I saw musical symbols beaming back at me: every known major religion and esoteric tradition in recorded history had embraced this science.... Here in fact, was the missing common factor I had long felt existed, that magical ingredient that had given religious movements the power to affect the minds and hearts of billions in such a profound and extraordinary way. They were all unerringly based on the principle of harmony, a harmony that it echoed in, literally, every single cell of our bodies, in our DNA and in the genetic code. This is, therefore, a natural harmony...[29]

In other words, the numbers of the Hermetic Code or Sacred Geometry appeared to be a legacy of a long-lost civilization, predating all the high civilizations of the classical era — the Sumerian, the Egyptian, the Mesopotamian, the Chinese — and the *essence* of that code was not only musical, but genetic. It was, as Hayes concluded, a "natural harmony," and DNA itself *was a natural resonator and transducer of it.*

2. The Well-Tempered Cosmos

But with the idea that it represents a "natural harmony," one also encounters yet another problem, a problem which reveals exactly how extraordinarily sophisticated the ancient Hermetic Code actually was. As I noted in my first book on alternative science and history, *The Giza Death Star*, the naturally occurring harmonic series is *not* what one encounters on a keyboard. To see what the problem is, we have to perform a simple experiment to demonstrate what physicists (and musicians!) know as the harmonic series.

Let us imagine one is sitting at the keyboard of an acoustic piano. Press down the note "C" silently with one hand, then strike the same note "C" an octave or two lower with the other hand. One will hear the open strings of the silently pressed note "C" vibrating sympathetically with the silently pressed note. This "Octave" is called the first harmonic or "overtone" of the fundamental "C" (the struck note). Now, repeat the experiment, only this time, press down the note "G" silently, but strike the note "C" an octave or two below it. Again, one will

---

29   Hayes, *The Hermetic Code in DNA,* pp. 19–20.

hear the note "G" vibrating sympathetically with the struck "C." G is the second overtone or harmonic of the fundamental "C," and, one notes that the interval has decreased, "G" being what musicians know as the interval of the *fifth* from C (counting up five notes from C, with C as "1," gives G as the fifth). If one keeps repeating the experiment, with the intervals growing smaller and smaller, then the next overtone will be a *fourth* up from G, which is another "C." The next interval is a major *third*, the note "E," then the next interval is a *minor* third, the note "G." But now we encounter the problem, for the *next interval, which occurs* **naturally**, is *not* present on the keyboard, for it is the interval *between the minor third and the major second*, or in other words a note lying *between B flat and A natural, or lying "in the cracks" on the keyboard between those two notes.*

This interval was known to the ancients, and in fact is called the "Pythagorean Comma," for the Pythagoreans were well aware of it. Yet, *as Hayes has pointed out, the analogy between ancient esoteric schools and modern science only works on a* **modern keyboard**. So what's going on here?

Briefly — and a thorough explanation of this relationship would require a book in its own right — what happened within Western music, *precisely as a result of the rediscovery of the esoteric tradition during the Renaissance,* was that that tradition was *applied to music* and to the tuning of musical instruments, particularly keyboards, so that the naturally occurring harmonic series was deliberately *tampered with,* and a slight mathematical adjustment was made to the notes of the musical scale, so that each note was exactly an equal interval apart, allowing *all notes to function as overtones of all other notes*, and thus allowing a piece of music to change keys during a piece to the most distant keys without having to stop and retune the whole instrument! This "tampered keyboard" ushered in the era of modern Western music, beginning with Vivaldi, Scarlatti, Rameau and J.S. Bach, who even celebrated its arrival by composing a piece of music called *The Well-Tempered Clavier,* or to put it bluntly, *"The Well-TAMPERED-WITH Keyboard."* Rather than each note now having its own unique harmonic series which did *not* overlap with other notes, each note now could function as any harmonic of any other note. *This, and not Maxwell's electromagnetic theory, was the first unification in physics, for now, rather than an infinite number of "notes" each with their own unique harmonic series, one had only 12 notes, each of which could function as harmonics of all the others.* The modern system of music had originated, in other words, with a slight adjustment in the system of *measurement* of musical intervals.

This little musical detour now exhibits a profound implication, for what it really implies is *that the ancients knew of this musical system, and, moreover, knew that it held some connection to DNA itself.* This is a profound indicator that the esoteric and Hermetic Code is a legacy of a very sophisticated scientific, and musical, culture.

Hayes is quick to appreciate the implications of these considerations in terms of modern quantum mechanics and its view of quantum reality as being a non-local, interconnected universe. Not surprisingly, the physicist whom he chooses as an example to illustrate this harmonic interconnectedness of systems in a very deep, non-local hyper-dimensional reality, is David Bohm!

> In an attempt to explain the principle of nonlocality and the idea of a vast web of interconnectedness permeating the whole universe, the University of London physicist David Bohm posited the existence of what he called quantum potential. He saw this as a new kind of energy field that, like gravity, pervades the whole universe, but whose influence does not weaken with distance.
>
> Bohm first recognized a possible indication of this quantum potential through his work on plasmas, gases comprising a high density of electrons and positive ions (atoms with a positive charge). He noticed that the electrons, once they were in plasma, began to act in concert, as if they were all part of a greater, interconnected whole. For example, if any impurities were present in the plasma, it would always realign itself and trap all foreign bodies in an exclusion zone — just as a living organism might encase poison in a boil. Bohm observed also a similar, orchestrated mass movement of electrons in metals and superconductors, with each one acting as if it "knew" what countless billions of others were about to do. According to Bohm, particles act in this way through the influence of the quantum potential, a subquantum force matrix that somehow coordinates the movement of the whole.
>
> It appears that when plasmas are rejecting impure substances and regenerating themselves, they look very similar to swirling masses of well-organized protoplasm. This curious "organic" quality led Bohm to comment that he often had the impression that the electron sea was, in a sense, "alive." He possibly did not intend this to be taken too literally, that the electron mass was living in the same way as an amoeba, but the evident highly coordinated symmetries of the plasma convinced him that the electrons were responding to one of many "intelligent" orders implicit in the fabric of the universe.[30]

Hayes comments on these observations and their implications, drawing attention to the "interlocking" nature of all phenomena via the property of resonance:

---

30   Hayes, *The Hermetic Code in DNA*, p. 93.

The whole universe is perpetually in motion and all wave/particles are continuously interacting and separating, which means that the nonlocal aspect of quantum systems in a general characteristic of nature. Clearly this represents, in the physical world, a harmony of the highest possible order. It is one thing to say that the universe is a harmonious entity because it is constructed entirely upon the eightfold chromodynamic and atomic matrices, but non-locality suggests that there exists a far deeper interconnecting harmony underlying all physical phenomena, where everything is resonating at the very same subquantum frequency, everything is "in tune" with every other thing,[31]

exactly as if it were one of the "notes" on a well-tempered cosmic keyboard, functioning as an overtone to all other notes in a vast and interconnected harmonic series.

And here, DNA enters the picture, for as Hayes has demonstrated, DNA would appear to be, in some sense, a "well-tempered" clavier for the whole system, or, exactly as the ancients understood mankind, a "microcosm." But is there any correlation between modern science and this ancient esoteric doctrine?

Indeed there is.

In my book *The Philosophers' Stone: Alchemy and the Secret Research for Exotic Matter*, I referred to a scientific paper by Freeman W. Cope entitled "Evidence from Activation Energies for Super-Conductive Tunneling in Biological Systems at Physiological Temperatures," a title guaranteed to make the eyes glaze over, unless one considers the profound implications hidden behind the dull scientific prose. But first, the dull scientific prose:

> In the present paper, evidence for another class of solid-state biological process is given. It is suggested that *single-electron tunneling between superconductive regions may rate-limit various nerve and growth processes. This implies that micro-regions of superconductivity exist in cells at physiological temperatures, which supports theoretical predictions of high temperature organic superconduction.*
>
> Superconduction is the passage of electron current without generation of heat and hence with zero electrical resistance. Such behavior has been observed only in inorganic materials and only at temperatures below approximately 20°K, although theory predicts

---

31   Ibid., p. 159.

that superconduction might occur in organic materials at room temperatures. The conduction of electrons *across interfaces* between adjacent superconductive layers behaves differently from current across ordinary solid junctions. Electron-tunneling currents across interfaces between superconductive layers or regions have been predicted and demonstrated to have a particular form of temperature dependence...

...Little...has suggested DNA as the sort of biological molecule along which electrons might superconduct...[32]

In *The Philosophers' Stone*, my focus was on the enormous implications of DNA superconductivity, but here, our interest is on the other aspect of the phenomenon noticed by Cope: electron *tunneling*.

The quantum tunneling phenomenon may be briefly understood by an illustration of what actually happens in quantum tunneling. Imagine one has an impermeable and impenetrable barrier, such as the wall of a biological cell, or a thick wall between two rooms, or even imagine the barrier as being the medium of space-time itself. Now, imagine a packet of waves approaching this barrier, like the interference pattern of two radar beams being bounced off the barrier and reflected back to the radar sets that generated the packet. But according to the wave equations of quantum mechanics, at the exact moment that this packet hits the barrier and is reflected *back* to the sets, a faint "*echo*" of that oncoming wave packet actually emerges on the *other* side of the barrier at the same instant, and travels away from the barrier; in short, the oncoming signal "tunnels" through the barrier — in almost wormhole-like fashion — and its "echo" emerges on the other side and travels away from the barrier, 180 degrees opposite of the reflected signal traveling back to our radar sets. And, as Cope notes in the section of the article cited above, DNA acts as exactly the sort of organic molecule that is able to accomplish this feat! It is as if an organic life form was a complex system being held together by superconductivity and quantum tunneling, a vast biological "non-local network" of tunneling nerve impulses. If so, then this might explain the basis of paranormal phenomena such as remote viewing, for DNA as such would interact *directly* with the information content within the field of the local physical medium.[33]

With this, then, we have the final component of that ancient technology

---

[32] Freeman W. Cope, "Evidence from Activation Energies for Super-conductive Tunneling in Biological Systems at Physiological Temperatures," *Physiological Chemistry and Physics 3* (1971), 403–410, pp. 403–404, emphasis added.

[33] It would, moreover, explain the fact that I first pointed out in my book *The Cosmic War: Interplanetary Warfare, Modern Physics, and Ancient Texts,* why the alleged high technology of the ancient gods appeared to be activated only in close proximity or actual physical contact with their owners. See chapters 8 and 9 of that work.

and science — the genetic and biological — and with it, we return once again to the banksters, and why they seem inordinately interested in learning its secrets.

D. The Ancient Contact: The Rothschild Nimrod Myth in a Wider Context

If DNA is such a resonator and transducer of the information content in the local field of the physical medium, it stands to reason that *some* specific genotypes will be more efficient resonators of it than others. We are told in countless ancient texts, from the Bible to the Sumerian epics to the Hindu epics of India to the legends of the Japanese "Yamato" peoples to the legends of Mesoamerican Indians, that at some point of time lost in the mists of "pre-history," the "gods" came down and mingled with men, siring children of human women. In many of these legends, moreover, these ancient "civilizing gods" were self-evidently white Caucasians; again, this is a common feature of such myths from Mesoamerica, to the South American Incans, and even ancient Japanese legends.

Whatever the truthfulness of these claims might be, we have seen that, since ancient times, an international banking class has consistently aligned itself with the temple, both with a view to giving their financial activities the sanction of the probity of the temple and thereby cloaking their actions, but also with a view to accessing the residues of that lost science and technology and potentially reconstructing and reconstituting it. And one such component, since the times of ancient Sumer and Babylonia, is surely their Sumerian epics' insistence that at least *some* of humanity descends from such "divine" ancestors and their intermarriages with humans. And given the laws of genetics, it is possible that, with a broad enough database, the outlines of such a genetic intermingling might eventually be discovered.

Such a database is, finally, within modern man's grasp, with the Human Genome Project. While many researchers point out the obvious and deadly potential that such a project entails for the design and engineering of race-specific bio-weapons, I am bold to suggest that there might be yet another hidden purpose, namely, to find and isolate those genotypes within humanity that point to a possible "extraterrestrial" connection. Again, the myth of the Rothschild descent from Nimrod — himself a product of such intermarriages if one consults the Sumerian records — suggests that one such banking dynasty is aware of this connection.

And indeed, as researchers Glen Yeadon and John Hawkins have noted, there is a deep connection between the Human Genome Project and the hidden hand of the Anglo-American international bankster class, and a deeper connection to the American eugenics movement with its own white-supremacist overtones:

> Currently, [the Cold Spring Harbor Laboratory] is a leader in the human genome project. While the genome project will undoubtedly provide many future medical benefits, it could equally provide weapons of mass destruction, such as bioweapons or even more evil genome-specific bioweapons. The *Project for the New American Century* [PNAC] describes genome-specific weapons as politically useful tools. PNAC serves as the blueprint for the George W. Bush administration, with many members closely associated with PNAC.
>
> With the Bush administration's disregard for human rights and the ban on nuclear testing, *it is cause for alarm to find Cold Spring Harbor firmly controlled by the same families involved in the American eugenics movement. Current directors William Gerry and Allen Dulles Jebsen are the grandsons of Averill Harriman and Allen Dulles, respectively. When such policies and organizations slip under the control of families like the Bushes and Rockefellers* they can be used as modern-day weapons of genocide.[34]

While the close association of "the same old families" with the genome project is a cause of concern, and for the same reasons Yeadon and Hawkins suggest, within the wider context of *this* book the close association of these families with the project calls forth concern for more than the reason they elicit, for it would appear that, once again, these families are trying their best, through all the vast network of foundations, grants, corporations and other fronts at their disposal, to learn and reconstruct as much of that ancient science, and perhaps of their own family history, as possible.

With DNA, we have the last component of the physics puzzle, the other pieces being, as was seen, the data of recurrent economic cycles, the peculiar correspondence of astrological alignments with economic depressions, the idea of a deep connection between the physics of the medium and the false alchemy of the banksters. We have seen the persistent efforts of some nations to free themselves from the influence of that class, and, in Nazi Germany's case, the attempt also to crash develop the paradigms of a new — and very old — physics. We have seen the persistent association of the banksters with religion, with the temple, and in turn, the persistent association of the ancient temple with astronomical alignments, with positioning on a global grid, and with the peculiar incorporation of "scalar physics" into the very design and dimensions of ancient temples.

So now, it is time to assemble the pieces together, and explain what's going on, and to offer a speculative explanation of *why* it's going on...

---

34 Glen Yeadon and John Hawkins, *The Nazi Hydra In America: Suppressed History of a Century* (Progressive Press, 2008), p. 173, emphasis added.

Nine

THE BANKSTERS' REAL BUSINESS
THE PATTERN OF WAR, SCARCITY, SUPPRESSION, SLAVERY AND MONOPOLIZATION

※

> *"In the course of the next several decades, a functioning structure of global cooperation, based on geopolitical realities, could thus emerge and gradually assume the mantle of the world's current "regent," which has for the time being assumed the burden of responsibility for world stability and peace. Geostrategic success in that cause would represent a fitting legacy of America's role as the first, only, and last truly global superpower."*
> —Zbigniew Brzezinski[1]

The fact that so many ancient temples heretofore surveyed show evidence not only of a profound association with the stars through their astrological alignments, and with the physics of the medium through their incorporation of sacred geometries and their location on an "earth grid" over nodal points of standing waves within local space, but also with international banking through the prominent association of moneychangers with those temples, is an indicator that at a deep and profound level — perhaps as a legacy of the Very High Civilization from which they sprang — these classical civilizations through their marriage of banking and astrology preserved the dim memory of a lost science that unified physics, economics, and finance.

---

1 Zbigniew Brzezinski, *The Grand Chessboard: American Primacy and Its Geostrategic Imperatives* (Basic Books, 1997), p. 215.

A. The Historical Patterns of Suppression
1. Of the Financial Alchemy of Money-as-Credit of the Nation

Moreover, we have presented evidence that this class existed in more or less unbroken continuity, both racially and ethnically, and, in terms of the methods and policies it has pursued throughout history. Additionally, suggestive allegations were presented that at least one member family of this class — the Rothschilds — trace their lineage directly back to one of the Sumerian elite, Nimrod.

In this, at least the methods surveyed have been consistent:

1) The basic methodology of the banksters, both in ancient and in modern times, has been to substitute one notion of money for another, and thereby to usurp the power of the state to create and issue money and substitute a private monopoly for its creation. In the former instance, the idea of money is that of a receipt for goods and services on the productive output of the state itself. In other words, money is "real" money and is a unit of information, of exchange, based on the creative activity of a whole state. It is thus issued by the state itself, and free of any interest-bearing debt. In the latter case, money is "false" money, or what we have called a *facsimile of* money, issued by a private monopoly and bearing debt interest, and circulated *as* money. The facsimile of money, in other words, is a debt-bearing note, and debt, under such a system, can only increase and never be repaid. In the first instance, it is to be noted, money represents the creation of *information,* i.e., units of exchange, *issued in proportion to the productive output of a state, and as such, the unit of exchange represents real information in the physical world.* In this instance, the connection between *finance and physics is direct and overt, and easily perceived.* It is therefore an analogue of the actual alchemical and transmutative properties, ultimately, of the physical medium itself. In the second instance, however, money represents almost *the exact opposite,* since it represents not production but an interest-bearing *debt,* a kind of "negative information" that can only and must inevitably grow at a rate exponentially faster than the productive output of the state. It bears, in other words, no relation to physical reality. It is a kind of false alchemy, a financial black hole into which the productive output of a state is inevitably sucked, and from which it *can never emerge, unless the very underlying idea of the facsimile of money is clearly repudiated and rejected.* The system of the facsimile of

money is, in other words, deliberately designed to enhance and enrich only those who control its issuance, and can never serve the broader public good because it is not *designed* to do so.

a) In the methods pursued to replace money by the facsimile of money, the bankster class, as we have seen, has invariably and throughout the centuries consistently followed a well-worn playbook, namely, the creation or counterfeiting of the real money of a state to undermine public confidence in it, or, in some cases, it has purchased the state's money issuance and taken it out of circulation, leaving only its own debt-bearing notes to circulate as a medium of exchange.

b) As we have seen from the historical examples presented in preceding pages, whenever a leader or a state has attempted to reject the principle of the facsimile of money, the international bankster class has mobilized those states it *does* control *through* that practice to make war against such "breakaway" states, or alternatively, the leaders of states advocating such policies of return to state-issued money as credit, are curiously murdered with a consistency that belies mere coincidence. In this respect, the following exemplars were mentioned in previous pages

[1] In modern times:

    [a] The manipulation of the collapse of the issuances of the Continental Congress during the 1780s by a return to the "gold" or "bullion" standard, which greatly contracted the American money supply, causing a depression, during a period when planetary alignments always seem to suggest economic collapse or contraction;

    [b] The assassination of President Abraham Lincoln for issuing debt-free Greenbacks to finance the northern effort during the American War Between the States, and the subsequent National currency act following a similar manipulated depression, which itself occurred during similar planetary alignments;

    [c] The assassination of President James Garfield mere months into his presidency, after he had suggested his willingness to return to state-created and -issued debt-free money;

    [d] The manipulation of the American money supply prior to and during the Great Depression, first by

vastly expanding credit, and then by vastly contracting the money supply, again at a juncture when planetary alignments indicated economic turmoil, and, as was seen, when Commerce Department data supplied by Edward Dewey indicated an inevitable cyclic downturn.

[e] Some researchers, as was seen in the previous pages, implied that the war against Nazi Germany had a secret purpose, for the banking elite of the West, having aided Adolf Hitler's and the Nazis' rise to power, were quickly alarmed by Nazi Germany's deliberate moves to restore state-issued debt-free money and by her efforts to explore alternative paradigms of physics that would lead to her energy independence and self-sufficiency;

[f] Finally, and as many researchers have noted, President John F. Kennedy was murdered a mere five months after his issuance of an executive order authorizing the U.S. Treasury to print $4,000,000,000 worth of debt-free United States Notes, bypassing the privately owned and controlled Federal Reserve Bank completely.

[2] Similarly, in ancient times, the following was noted:

[a] the policies of Rome vis-à-vis its trading partners in the East regarding the value of gold and silver bullion strongly indicated manipulation at both ends of the trading routes, manipulation ultimately favorable to any class controlling bullion supplies;

[b] Similar policies of issuance of false receipts or counterfeits of government currencies were noted, with the ultimate issuance of debt-bearing bullion-backed notes of debt by a private monopoly, particularly in the case of ancient Babylonia.

[c] In the case of Sparta, it was also noted that it alone of all the ancient Greek city-states resisted this financial policy, a fact which led ultimately to the Peloponnesian War and to the eventual succumbing of Sparta to these policies and the class behind them.

2. Of the Physical Alchemy of Energy from the Medium

2) Similarly, both in ancient times and in modern ones, we have noted a persistent association of the banking class with the following things:
   a) In ancient times, with religion, i.e., the *temple*. The temple, in ancient times, was in turn associated with the following things:
      [1] With astronomy and astrology, i.e., with the "science" of forecasting human events;
      [2] With sacred geometries, embodied both in the dimensions of the structure, and in their placements and alignments on an earth grid;
      [3] With the issuance of *money*, both in the forms of real money, and its facsimile as interest-bearing debt;
      [4] With slavery.
   b) In modern times, with not only religion, but with science and scientists determined to break *out* of the paradigm of energy scarcity being based on non-renewable resources, and who were seeking to develop new physics paradigms, with new energy technologies.
   c) And in modern times, we saw evidence of corporate and banking interest in the vast database indicating a correlation and correspondence between economic cycles, and cycles of purely *physical* activity, cycles that, in Dewey's estimation, were inevitable *because they were due to deeper underlying physical reasons and not simply the result of policy or aggregate human action.*

But the question, highlighted in this fashion, only remains: *Why* are they persistently associated with these things? Why *does* one find banksters manipulating such drastic curtailings of money supply when both cyclic data and astrological lore would indicate that such downturns were inevitable?

The answer to these questions requires that we now assemble all the *physics* pieces of the puzzle, outlined in the preceding pages, into a single catalogue, determine their underlying principles, and then correlate these with the above patterns of bankster activities and policies. Only then does the disturbing answer and picture of their motivations and possible agenda emerge.

B. The Physics, The Financial Alchemy, and the Banksters
1. Assembling the Pieces

Looking back over the assembled physics and financial data of previous chapters, the following salient features emerge:

1) According to Dewey and Dakin and the Foundation for the Study of Cycles, economic activity occurs in cycles of boom and bust, and these cycles are exact analogues to physical *longitudinal* waves of sound, i.e., of compression and rarefaction. As such, these cycles are in a certain sense inevitable and impervious to human action, but the overall downward and upward *trend* of a cycle can be dampened or exacerbated by human policy and activity;
2) Dewey and Dakin also posit that a deeper physics may lie behind such economic cycles, a fact loosely corroborated by Nelson's study of the variations in radio signal propagation and planetary alignments, and by Gover's study of American depressions which seem to occur near astrological alignments called grand crosses. It was also noted that the grand cross pattern mentioned by Gover also closely corresponded to the type of planetary alignments noted by Nelson in his RCA study;
3) All of this was further corroborated by the Global Scaling Theory of Dr. Harmut Müller, who noticed that physical objects themselves — including human social and demographic activity — clustered around certain nodal points of standing waves in logarithmic space. Müller went on to posit that these nodal points may indicate a real physics of the physical medium, i.e., that such clustering was due to the interference of longitudinal waves in the physical medium itself. On this view, Müller's theory is very close to the views of ancient alchemy which posits that the physical medium is a *transmutative* medium, creating information in its differentiations which in turn become the physical objects of creation. As was noted, Müller's theory came to the attention of the German Institute for Space Energy Research, which applied his theories to note the deep physics behind human social action and organization;
4) Similar views of matter *as a template or interference pattern of such waves* were expressed by quantum and plasma physicist David Bohm;
5) Since in these views matter is the result of a template or interference pattern of such waves in the medium, large masses such as

planets and stars are *natural resonators* of such waves. Stars in particular are resonators of such waves, since, as *rotating balls of plasma* (which in turn, according to David Bohm, exhibit phenomena of "self-organization" and response to the "intelligent" instructions of the physical medium) they act as efficient natural resonators of the medium. As Richard C. Hoagland noted, it appears that the ancient societies, with their emphasis on stars as "portals" or "gates" to higher dimensions, preserved a legacy of this deep physics;

6) As natural resonators of these standing waves in the medium, it was also noted that planets set up within themselves a gridwork or template of such wave interference patterns on their surfaces, which, it was posited, may be the physical basis for the evidence of an "earth grid," and of the placement of ancient sites and temples over these nodal points. A similar claim was made by Tesla for his wireless transmission of power facility on Long Island called Wardenclyffe. According to Tesla, in order for his system to work, it was necessary for it to "grip the earth";

7) As Dr. Konstantin Meyl also noted, many such temples themselves appear to be constructed in their dimensions to be natural resonators of such waves. Meyl posited that such temples actually functioned as a means of instantaneous radio communications by means of scalar waves, using the sensitive membranes of slaughtered animals to register and measure the pulses of such waves; and finally,

8) As noted by British researcher Hayes, DNA appears to be designed along the same lines and principles of sacred geometry and the hermetic code. As such, DNA itself would seem to be a natural resonator of such waves.

2. *The Underlying Principle*

Examining each of these eight points, an obvious underlying physical principle of action is implied, which may be reduced to three components:

1) All action within the physical medium of space-time itself is constituted of longitudinal waves of compression and rarefaction within it;

2) These in turn give rise to matter, which in the form of planets and more importantly, stars (or *rotating plasma*), are natural transducers of such waves; and,

3) A system of such masses — i.e., a solar system — will, in its constantly changing dynamics, both resonate to and in turn give off an ever-changing dynamic and interference pattern or template of such waves, forming, perhaps, the ultimate and now lost scientific basis of astrology. In this, since DNA would seem to be a natural resonator of such interference patterns or templates, through long process of observation — the precise claim made for astrology in ancient Sumeria — it may have been possible to predict human response, via DNA, to such patterns *in the aggregate,* as Dewey and Dakin suggested, and as the influx of physicists with their analytical methods into economics and finance in the latter half of the twentieth century implies.[2]

a. The Reasons for the Banksters' Ancient Associations with the Temple

We are now, at last, in a position to see why at least in ancient times one finds the persistent pattern of the presence of "bullion brokers" or "banksters" with the temple, for with the astrological and oracular preoccupations of those temples, this would give such a class immediate access to

1) *prediction,* i.e., advance knowledge of cycles of human activity and emotional responses;
2) *communication,* for if Meyl's hypothesis should be borne out by subsequent scientific investigation and experiment, access to the temple would have given them access to a means of virtually instantaneous communication across a great part of the world's surface;
3) This in turn would have thus afforded them a means to concerted and coordinated activity and *social manipulation,* a coordination we saw was strongly suggested by the bullion policies of ancient Rome vis-à-vis the Far East; and finally,
4) If one assumes the truthfulness of the scenario of an ancient and cosmic war outlined at the beginning of chapter one, access to the residues of this lost science via a close association with the ancient temple would have also given the banksters the means, ultimately,

---

2 This would, perforce, imply an astonishing thing, namely, that Dr. Li's Gaussian copula formula failed for yet *another* reason than those advanced in the Prologue, and that the failure was due to not recognizing a deeper physics in play. The data compiled by the Foundation for the Study of Cycles, as well as other data presented here, would seem to indicate also that at least *some* in the international finance "community" are *fully aware* of this hypothesis, and plan their actions accordingly.

to *recover and reconstruct* the technologies of hegemony and mass destruction that are implied by that scenario, technologies which in turn are based on exactly the same *type* or paradigm of physics.

It will be evident from the foregoing list that this pattern has persisted into modern times; witness only the close association of Morgan with Tesla, and Morgan's eventual *suppression* of Tesla. There are countless other such examples, of course: ITT's suppression of Philo Farnsworth's successful plasma fusion experiments, the suppression of Dr. Ronald Richter's "fusion" experiments in Argentina,[3] and on and on the list could go. The same pattern holds, as well, during the "interim" period *between* ancient and modern times, for during that period, one finds the persistent interest of the royal houses in Europe in *alchemy,* the ability to transmute base metals into gold. The reasons are now evident, for the private issuance of the facsimile of money as an interest-bearing debt note took from them the most ancient prerogative of governments and kings: the issuance of debt-free money on the productive output of the state.

The motivations are clear enough from the *physics,* for at one and the same time the banksters were trying to recover that ancient physics, and with it, the genuine financial alchemy of state-created money. The banksters, for their part, are also found during the interim period, and at both the modern and ancient poles of the story, resorting to a kind of false alchemy of debt, a kind of "negative financial information," while they are simultaneously and very quietly funding certain scientists who hold the promise of unlocking that ancient physics. Witness only Isaac Newton's preoccupations with alchemy, his membership in the Masonically controlled and inspired Royal Society, and the close connection of all these parties with the Dutch bankers that quietly and quickly assumed control of England's money.[4]

3. The Possible Agenda

The careful reader will have noticed an immediate implication of the previous considerations, and that is that, as far back as ancient times, the persistent agenda of the bankster class, the international "bullion brokers," has been coordinated action that aimed at a global reach, with ever larger and larger empires being created, provoking ever more wars, and ever more business for themselves. And thus we end where we began, in modern times, and the persistent and consistent evidence that this class is truly after a global

---

3 See my *Nazi International,* chapters 8 and 9.
4 For the complete story of how this was accomplished, see Alexander Del Mar's *Barbara Villiers: A History of Monetary Crimes.*

hegemony. As David Astle pointed out and as was reviewed in chapter five, this hegemony quickly, and very early in its history, realized that all ancient pantheons and myths were more or less identical, and thus that religion itself not only held profound clues to the reconstruction of the lost science they sought to recover, but that it could function as a powerful tool of social manipulation to assist in that agenda. Not for nothing does one find a Rockefeller dynasty supporting the work and agenda of the World Council of Churches from its inception, or of various seminaries with a "suitable" social agenda, nor a Rothschild dynasty sponsoring various similar work. In this, "monotheism" really masks as a front for their own designs, with the claims of the three great monotheistic religions — Christianity, Judaism, and Islam — seemingly tailor-made for the manipulation and exploitation of conflict.

But to what end all this machination?

While many have guessed at the motivations for this vast and ancient conspiracy, no one seems to have approached anything like a final answer. I certainly cannot claim to have done so either. But an answer *does* suggest itself from the preceding pages, namely, that they are indeed trying to reconstruct a lost mythical past: a global "golden age" with a supremely sophisticated science with which they can dominate and subjugate the earth. But to reconstruct it, on the scale required and implied by their enterprise itself, will require that virtually the entire planet and its resources must be at their disposal.

What they intend to do after that is beyond the scope and purpose of this essay, but an answer does suggest itself, for if, as was seen, at least one of these banking dynasties — the Rothschilds — are alleged to trace their lineage back to Nimrod, half-human half-divine offspring of the "gods" who once descended to earth and sired children of human mothers, then perhaps they seek, ultimately, to return to the stars.

But....

...tracing that lineage, and that motivation, is a whole *other* story...

For *this* story, however, the lesson is clear: the false alchemy of the banksters and their issuance of the facsimile of money can only mean that money is based on *scarcity and debt,* on non-renewable energy sources, and on a paradigm of physics and energy, and a system that must ultimately consume itself in endless slavery and wars.

It is high time to have done with them.

BIBLIOGRAPHY

No Author. "Econophysics," en.wikipedia.org/wiki/Econophysics

No author. "Edward R. Dewey," Wikipedia, en.wikipedia.org/wiki/Foundation _for _the_Study_of_Cycles

No author. "Hermann Josef Abs," moversandshakersofthesmom.blogspot.com/2008/08/hermann-abs.html

No author. "RCA Astrology," *Time,* Monday, April 16, 1951, www.time.com/magazine/article/0,9171,814720,00.html

No author. www.cycleslibrary.org/synchronies/

No author. "Tesla's Wireless Torpedo: Inventor Says He Did Show that it Worked Perfectly," *New York Times,* March 19, 1907.

"Mr. Tesla's Invention: How the Electrician's Lamp of Aladdin May Construct New Worlds," *New York Times,* April 21, 1908.

No Author. "Earth Grid Theories," www.vortexmaps.com/devils-triangle.php

Alfvén, Hannes. "Existence of Electromagnetic-Hydrodynamic Waves," *Nature,* No. 3805, October 3, 1942, pp. 405–406.

Alfvén, Hannes. "On Hierarchical Cosmology," *Astrophysics and Space Science,* Vol. 89. Boston: D. Reidel, 1983.

Astle, David. *The Babylonian Woe,* ellhn.eee.gr. No date.

Bohm, David. *Wholeness and the Implicate Order.* London: Routledge, 1995. ISBN 0-415-11966-9.

Brzezinski, Zbigniew. *The Grand Chessboard: American Primacy and Its Geostrategic Imperatives.* Basic Books, 1997. ISBN 978-0-465-02726-2.

Brown, Ellen Hodgson, J.D. *Web of Debt: The Shocking Truth about Our Money System and How We Can Break Free.* Baton Rouge: Third Millennium Press, 2008. ISBN 978-0-9795608-2-8.

Childress, David Hatcher, ed. *The Fantastic Inventions of Nikola Tesla.* Kempton, Illinois: Adventures Unlimited Press, 1993. ISBN 0-932813-19-4.

Cope, Freeman A. "Evidence from Activation Energies for Super-conductive Tunneling in Biological Systems at Physiological Temperatures." *Physiological Chemistry and Physics 3* (1971), 403–410.

Dawson, Christopher. *The Age of the Gods.* London, 1928.

Del Mar, Alexander, *Barbara Villiers: A History of Monetary Crimes.* Honolulu: University Press of the Pacific, 2004. ISBN 1-4102-1102-9.

Del Mar, Alexander. *A History of Money in Ancient Countries from the Earliest Times to the Present.* Kessinger Publications reprint of George Bell and Sons, 1885. ISBN 0-7661-9024-2.

Del Mar, Alexander. *A History of Monetary Systems.* Honolulu: University Press of the Pacific, 2000. ISBN 0-89875-062-8.

Del Mar, Alexander. *A History of the Precious Metals from the Earliest Times to the Present.* Elibron Classics, 2005. ISBN 1-4021-7302-4.

Desborough, Brian. *They Cast No Shadows: A Collection of Essays on the Illuminati, Revisionist History, and Suppressed Technologies.* San Jose, California: Writers Club Press, 2002.

Dewey, Edward R., and Edwin F. Dakin, *Cycles: The Science of Prediction.* Henry Holt and Company, 1947. (Reprinted by Kessinger Publishing.) ISBN 1436710219.

Estulin, Daniel. *The True Story of the Bilderberg Group.* Walterville, Oregon: TrineDay LLC, 2007. ISBN 978-0-9777953-4-5.

Farrell, Joseph P. *The Cosmic War: Interplanetary Warfare, Modern Physics, and Ancient Texts.* Kempton, Illinois: Adventures Unlimited Press, 2006. ISBN 978-1931882750.

Farrell, Joseph P. *The Nazi International: The Nazis' Secret Postwar Plan to Control Finance, Conflict, Physics, and Space.* Kempton, Illinois: Adventures Unlimited Press, 2008. ISBN 978-1931882934.

Farrell, Joseph P. *The Philosophers' Stone: Alchemy and the Secret Research for Exotic Matter: The American "Gold," The Soviet "Mercury," and The Nazi "Serum."* Port Townsend, Washington: Feral House, 2009. ISBN 978-1932595406.

Frank, Tenney. "An Economic History of Rome." *The American Historical Review,* Vol. 21, July 21, 1916.

Gover, Robert. *Time and Money: The Economy and the Planets.* Titusville, New Jersey: Hopewell Publications, LLC, 2005. ISBN 0-9726906-8-9.

Hayes, Michael. *The Hermetic Code in DNA: The Sacred Principles in the Ordering of the Universe.* Rochester, Vermont: Inner Traditions, 2008. ISBN 978-159477218-4.

Heri, Sesh. *The Handprint of Atlas: The Artificial Axis of the Earth and How it Shaped Human Destiny.* Highland, California: Corvos Books, Lost Continent Library Publishing Co., 2008.

Hoagland, Richard C. *Hoagland's Mars: Vol. 2: The United Nations Briefing:* (UFO TV DVD).

Kunz, George Frederick. *The Curious Lore of Precious Stones.* Dover, 1971. ISBN 0-486-22227-6.

LaViolette, Paul A., Ph.D. *Secrets of Antigravity Propulsion: Tesla, UFOs, and Classified Aerospace Technology.* Rochester, Vermont: Bear & Company, 2008. ISBN 978-159143078-0.

Lerner, Eric J. *The Big Bang Never Happened.* New York: Vintage Books, 1992.

Lockyer, J. Norman. *The Dawn of Astronomy: A Study of Temple Worship and Mythology of the Ancient Egyptians.* Mineola, New York: Dover Publications. ISBN 0-486-45012-0.

Manning, Paul. *Martin Bormann: Nazi in Exile.* Lyle Stuart, Inc., 1981.

Marrs, Jim. *The Rise of the Fourth Reich: The Secret Societies that Threaten to Take Over America.* New York: Wm. Morrow (HarperCollins), 2008. ISBN 978-0-06-124558-9.

Meyl, Konstantin, Prof. Dr. *Scalar Waves.* Villingen-Schwenningen, Germany, 2003. ISBN 3-9802-542-4-0.

Müller, Dr. Harmut. "An Introduction to Global Scaling Theory." *Nexus,* Vol. 11, No. 5, September-October 2004.

Nelson, J.H. "Planetary Position Effect on Short-Wave Signal Quality." *Electrical Engineering,* May 1952.

Nelson, J.H. "Shortwave Radio Propagation Correlation With Planetary Positions." Conference paper presented to the AIEE Subcommittee on Energy Sources, AIEE General Winter Meeting, January 1952.

Nichelson, Oliver. "Tesla Wireless Power Transmitter and the Tunguska Explosion of 1908," prometheus.al.ru/english/phisik/onichelson/tunguska

Perkins, John. *Confessions of an Economic Hit Man.* New York: Plume Books (Penguin Group), 2006. ISBN 978-0-452-28708-2.

Quigley, Carroll. *The Anglo-American Establishment: From Rhodes to Cliveden.* San Pedro, California: GSG and Associates, Publishers, 1981. ISBN 0-945001-01-0

Quigley, Carroll: *Tragedy and Hope: A History of the World In Our Time.* Los Angeles: Wm. Morrison, 1974. ISBN 0913022-14-4.

Salmon, Felix. "A Formula for Disaster." *Wired,* March 2009.

Springmeier, Fritz. *Bloodlines of the Illuminati.* Ambassador House, 2002. ISBN 0-9663533-2-3.

Stevens, Henry. *Hitler's Suppressed and Still-Secret Weapons, Science, and Technology.* Kempton, Illinois: Adventures Unlimited Press, 2005.

Tarpley, Webster. "Lord Palmerston's Multicultural Human Zoo." www.mail-archive.com/ctrl@listserv.aol.com/msg17357.html

Tarpley, Webster. "Venice's War Against Western Civilization." www.schillerinstitute.org/fid_91-96/952_venice.html

Tesla, Nikola. "Tesla's New Device Like Bolts of Thor: He Seeks to Patent a Wireless Engine for Destroying Navies by Pulling a Lever; to Shatter Armies Also." *New York Times,* Dec. 8, 1915.

Tomes, Ray. "Towards a Unified Theory of Cycles." Paper presented at The Foundation for the Study of Cycles Conference Proceedings, February 1990, www.cyclesresearchinstitute.org

Weidner, Jay and Bridges, Vincent. *The Mysteries of the Great Cross of Hendaye: Alchemy and the End of Time.* Rochester, Vermont: Destiny Books, 2003.

FERAL HOUSE